# SHIP'S LOG

Designed By:
## JOHN P. KAUFMAN

Executive Editor: E. Robert "Bob" Lollo

BRISTOL FASHION PUBLICATIONS, INC.
Harrisburg, Pennsylvania

Published by Bristol Fashion Publications, Inc.

ISBN: 1-892216-08-6

# Horn Signals

Danger - 5 Short Blasts
Bridge Opening - 1 Long-1 Short Blast
Lock Opening - 2 Long-2 Short Blasts
Passing - 2 Blasts = Pass On Port - 1 Blast = Pass On Starboard

# How To Issue A Distress Call

Be certain the radio is turned on.

Press microphone button and repeat: Mayday-Mayday-Mayday.

Say: This is the vessel "*State Name of Boat*". Repeat three times.

Say: Mayday. My position is (give latitude and longitude from GPS or buoys in the area).

Say: The condition of the boat and/or crew and number of crew.

Say: We require. (State the type of help needed).

Say: *Boat Name* is a *Year, Hull Material, Type of Boat. Color Of Boat and Trim.*

Say: I am standing by on channel 16 for assistance.

Release the microphone button and listen to the radio.

If there is no answer in 10 seconds repeat the entire above message.

Continue this procedure until an answer is received or as long as it is safely possible to remain on the vessel.

# BEFORE LEAVING BOAT UNATTENDED

Flush and treat head

Lock lazarette

Turn off refrigerator

Turn off hot water

Turn off A/C or heat

Adjust heater to prevent freezing

Set charger

Set batteries

Check lines

All canvas down

Life vests secure

Electronics

Flags in

All keys

Lock all windows

Lock all hatches

Fishing gear

Lock cabin door

Garbage to dumpster

Shore water  or pressure pump off

Seacocks closed

# SHIP'S LOG

DATE_____CREW_____GUESTS_____PAGE_____

DEPARTURE TIME_____DESTINATION_____ARRIVAL TIME_____

| TIME | ENGINE HOURS | PILOT INT'L | POSITION | HEADING | RPM | SPEED | ENGINE TEMP | AMPS | OIL | FUEL TANK | BARO | TEMP | WIND |
|------|------|------|------|------|------|------|------|------|------|------|------|------|------|
|  |  |  |  |  |  |  |  |  |  |  |  |  |  |
|  |  |  |  |  |  |  |  |  |  |  |  |  |  |
|  |  |  |  |  |  |  |  |  |  |  |  |  |  |
|  |  |  |  |  |  |  |  |  |  |  |  |  |  |
|  |  |  |  |  |  |  |  |  |  |  |  |  |  |
|  |  |  |  |  |  |  |  |  |  |  |  |  |  |
|  |  |  |  |  |  |  |  |  |  |  |  |  |  |
|  |  |  |  |  |  |  |  |  |  |  |  |  |  |
|  |  |  |  |  |  |  |  |  |  |  |  |  |  |
|  |  |  |  |  |  |  |  |  |  |  |  |  |  |
|  |  |  |  |  |  |  |  |  |  |  |  |  |  |
|  |  |  |  |  |  |  |  |  |  |  |  |  |  |

## REMARKS & HAPPENINGS

# SHIP'S LOG

DATE_____CREW_____GUESTS_____PAGE_____

DEPARTURE TIME_____DESTINATION_____ARRIVAL TIME_____

| TIME | ENGINE HOURS | PILOT INT'L | POSITION | HEADING | RPM | SPEED | ENGINE TEMP | AMPS | OIL | FUEL TANK | BARO | TEMP | WIND |
|------|--------------|-------------|----------|---------|-----|-------|-------------|------|-----|-----------|------|------|------|
|      |              |             |          |         |     |       |             |      |     |           |      |      |      |
|      |              |             |          |         |     |       |             |      |     |           |      |      |      |
|      |              |             |          |         |     |       |             |      |     |           |      |      |      |
|      |              |             |          |         |     |       |             |      |     |           |      |      |      |
|      |              |             |          |         |     |       |             |      |     |           |      |      |      |
|      |              |             |          |         |     |       |             |      |     |           |      |      |      |
|      |              |             |          |         |     |       |             |      |     |           |      |      |      |
|      |              |             |          |         |     |       |             |      |     |           |      |      |      |
|      |              |             |          |         |     |       |             |      |     |           |      |      |      |
|      |              |             |          |         |     |       |             |      |     |           |      |      |      |
|      |              |             |          |         |     |       |             |      |     |           |      |      |      |
|      |              |             |          |         |     |       |             |      |     |           |      |      |      |
|      |              |             |          |         |     |       |             |      |     |           |      |      |      |
|      |              |             |          |         |     |       |             |      |     |           |      |      |      |
|      |              |             |          |         |     |       |             |      |     |           |      |      |      |
|      |              |             |          |         |     |       |             |      |     |           |      |      |      |

## REMARKS & HAPPENINGS

# SHIP'S LOG

DATE_____CREW_____GUESTS_____PAGE_____

DEPARTURE TIME_____DESTINATION_____ARRIVAL TIME_____

| TIME | ENGINE HOURS | PILOT INT'L | POSITION | HEADING | RPM | SPEED | ENGINE TEMP | AMPS | OIL | FUEL TANK | BARO | TEMP | WIND |
|------|------|------|------|------|------|------|------|------|------|------|------|------|------|
|      |      |      |      |      |      |      |      |      |      |      |      |      |      |
|      |      |      |      |      |      |      |      |      |      |      |      |      |      |
|      |      |      |      |      |      |      |      |      |      |      |      |      |      |
|      |      |      |      |      |      |      |      |      |      |      |      |      |      |
|      |      |      |      |      |      |      |      |      |      |      |      |      |      |
|      |      |      |      |      |      |      |      |      |      |      |      |      |      |
|      |      |      |      |      |      |      |      |      |      |      |      |      |      |
|      |      |      |      |      |      |      |      |      |      |      |      |      |      |
|      |      |      |      |      |      |      |      |      |      |      |      |      |      |
|      |      |      |      |      |      |      |      |      |      |      |      |      |      |
|      |      |      |      |      |      |      |      |      |      |      |      |      |      |
|      |      |      |      |      |      |      |      |      |      |      |      |      |      |
|      |      |      |      |      |      |      |      |      |      |      |      |      |      |
|      |      |      |      |      |      |      |      |      |      |      |      |      |      |

## REMARKS & HAPPENINGS

# SHIP'S LOG

DATE_____CREW_____GUESTS_____PAGE_____

DEPARTURE TIME_____DESTINATION_____ARRIVAL TIME_____

| TIME | ENGINE HOURS | PILOT INT'L | POSITION | HEADING | RPM | SPEED | ENGINE TEMP | AMPS | OIL | FUEL TANK | BARO | TEMP | WIND |
|------|------|------|------|------|------|------|------|------|------|------|------|------|------|
| | | | | | | | | | | | | | |
| | | | | | | | | | | | | | |
| | | | | | | | | | | | | | |
| | | | | | | | | | | | | | |
| | | | | | | | | | | | | | |
| | | | | | | | | | | | | | |
| | | | | | | | | | | | | | |
| | | | | | | | | | | | | | |
| | | | | | | | | | | | | | |
| | | | | | | | | | | | | | |
| | | | | | | | | | | | | | |
| | | | | | | | | | | | | | |
| | | | | | | | | | | | | | |
| | | | | | | | | | | | | | |
| | | | | | | | | | | | | | |
| | | | | | | | | | | | | | |
| | | | | | | | | | | | | | |
| | | | | | | | | | | | | | |

## REMARKS & HAPPENINGS

# SHIP'S LOG

DATE_____CREW_____GUESTS_____PAGE_____

DEPARTURE TIME_____DESTINATION_____ARRIVAL TIME_____

| TIME | ENGINE HOURS | PILOT INT'L | POSITION | HEADING | RPM | SPEED | ENGINE TEMP | AMPS | OIL | FUEL TANK | BARO | TEMP | WIND |
|------|------|------|------|------|------|------|------|------|------|------|------|------|------|
|  |  |  |  |  |  |  |  |  |  |  |  |  |  |
|  |  |  |  |  |  |  |  |  |  |  |  |  |  |
|  |  |  |  |  |  |  |  |  |  |  |  |  |  |
|  |  |  |  |  |  |  |  |  |  |  |  |  |  |
|  |  |  |  |  |  |  |  |  |  |  |  |  |  |
|  |  |  |  |  |  |  |  |  |  |  |  |  |  |
|  |  |  |  |  |  |  |  |  |  |  |  |  |  |
|  |  |  |  |  |  |  |  |  |  |  |  |  |  |
|  |  |  |  |  |  |  |  |  |  |  |  |  |  |
|  |  |  |  |  |  |  |  |  |  |  |  |  |  |
|  |  |  |  |  |  |  |  |  |  |  |  |  |  |
|  |  |  |  |  |  |  |  |  |  |  |  |  |  |
|  |  |  |  |  |  |  |  |  |  |  |  |  |  |
|  |  |  |  |  |  |  |  |  |  |  |  |  |  |
|  |  |  |  |  |  |  |  |  |  |  |  |  |  |

## REMARKS & HAPPENINGS

# SHIP'S LOG

DATE_____CREW_____GUESTS_____PAGE_____

DEPARTURE TIME_____DESTINATION_____ARRIVAL TIME_____

| TIME | ENGINE HOURS | PILOT INT'L | POSITION | HEADING | RPM | SPEED | ENGINE TEMP | AMPS | OIL | FUEL TANK | BARO | TEMP | WIND |
|------|------|------|------|------|------|------|------|------|------|------|------|------|------|
|  |  |  |  |  |  |  |  |  |  |  |  |  |  |
|  |  |  |  |  |  |  |  |  |  |  |  |  |  |
|  |  |  |  |  |  |  |  |  |  |  |  |  |  |
|  |  |  |  |  |  |  |  |  |  |  |  |  |  |
|  |  |  |  |  |  |  |  |  |  |  |  |  |  |
|  |  |  |  |  |  |  |  |  |  |  |  |  |  |
|  |  |  |  |  |  |  |  |  |  |  |  |  |  |
|  |  |  |  |  |  |  |  |  |  |  |  |  |  |
|  |  |  |  |  |  |  |  |  |  |  |  |  |  |
|  |  |  |  |  |  |  |  |  |  |  |  |  |  |
|  |  |  |  |  |  |  |  |  |  |  |  |  |  |
|  |  |  |  |  |  |  |  |  |  |  |  |  |  |
|  |  |  |  |  |  |  |  |  |  |  |  |  |  |

## REMARKS & HAPPENINGS

# SHIP'S LOG

DATE_____CREW_____GUESTS_____PAGE_____

DEPARTURE TIME_____DESTINATION_____ARRIVAL TIME_____

| TIME | ENGINE HOURS | PILOT INT'L | POSITION | HEADING | RPM | SPEED | ENGINE TEMP | AMPS | OIL | FUEL TANK | BARO | TEMP | WIND |
|------|------|------|------|------|------|------|------|------|------|------|------|------|------|
| | | | | | | | | | | | | | |
| | | | | | | | | | | | | | |
| | | | | | | | | | | | | | |
| | | | | | | | | | | | | | |
| | | | | | | | | | | | | | |
| | | | | | | | | | | | | | |
| | | | | | | | | | | | | | |
| | | | | | | | | | | | | | |
| | | | | | | | | | | | | | |
| | | | | | | | | | | | | | |
| | | | | | | | | | | | | | |
| | | | | | | | | | | | | | |
| | | | | | | | | | | | | | |

## REMARKS & HAPPENINGS

# SHIP'S LOG

DATE_____CREW_____GUESTS_____PAGE_____

DEPARTURE TIME_____DESTINATION_____ARRIVAL TIME_____

| TIME | ENGINE HOURS | PILOT INT'L | POSITION | HEADING | RPM | SPEED | ENGINE TEMP | AMPS | OIL | FUEL TANK | BARO | TEMP | WIND |
|------|------|------|------|------|------|------|------|------|------|------|------|------|------|
|  |  |  |  |  |  |  |  |  |  |  |  |  |  |
|  |  |  |  |  |  |  |  |  |  |  |  |  |  |
|  |  |  |  |  |  |  |  |  |  |  |  |  |  |
|  |  |  |  |  |  |  |  |  |  |  |  |  |  |
|  |  |  |  |  |  |  |  |  |  |  |  |  |  |
|  |  |  |  |  |  |  |  |  |  |  |  |  |  |
|  |  |  |  |  |  |  |  |  |  |  |  |  |  |
|  |  |  |  |  |  |  |  |  |  |  |  |  |  |
|  |  |  |  |  |  |  |  |  |  |  |  |  |  |
|  |  |  |  |  |  |  |  |  |  |  |  |  |  |
|  |  |  |  |  |  |  |  |  |  |  |  |  |  |
|  |  |  |  |  |  |  |  |  |  |  |  |  |  |
|  |  |  |  |  |  |  |  |  |  |  |  |  |  |
|  |  |  |  |  |  |  |  |  |  |  |  |  |  |

## REMARKS & HAPPENINGS

# SHIP'S LOG

DATE_____CREW_____GUESTS_____PAGE_____

DEPARTURE TIME_____DESTINATION_____ARRIVAL TIME_____

| TIME | ENGINE HOURS | PILOT INT'L | POSITION | HEADING | RPM | SPEED | ENGINE TEMP | AMPS | OIL | FUEL TANK | BARO | TEMP | WIND |
|------|------|------|------|------|------|------|------|------|------|------|------|------|------|
| | | | | | | | | | | | | | |
| | | | | | | | | | | | | | |
| | | | | | | | | | | | | | |
| | | | | | | | | | | | | | |
| | | | | | | | | | | | | | |
| | | | | | | | | | | | | | |
| | | | | | | | | | | | | | |
| | | | | | | | | | | | | | |
| | | | | | | | | | | | | | |
| | | | | | | | | | | | | | |
| | | | | | | | | | | | | | |
| | | | | | | | | | | | | | |
| | | | | | | | | | | | | | |
| | | | | | | | | | | | | | |
| | | | | | | | | | | | | | |
| | | | | | | | | | | | | | |
| | | | | | | | | | | | | | |

## REMARKS & HAPPENINGS

# SHIP'S LOG

DATE_____CREW_____GUESTS_____PAGE_____

DEPARTURE TIME_____DESTINATION_____ARRIVAL TIME_____

| TIME | ENGINE HOURS | PILOT INT'L | POSITION | HEADING | RPM | SPEED | ENGINE TEMP | AMPS | OIL | FUEL TANK | BARO | TEMP | WIND |
|------|------|------|------|------|------|------|------|------|------|------|------|------|------|
|  |  |  |  |  |  |  |  |  |  |  |  |  |  |
|  |  |  |  |  |  |  |  |  |  |  |  |  |  |
|  |  |  |  |  |  |  |  |  |  |  |  |  |  |
|  |  |  |  |  |  |  |  |  |  |  |  |  |  |
|  |  |  |  |  |  |  |  |  |  |  |  |  |  |
|  |  |  |  |  |  |  |  |  |  |  |  |  |  |
|  |  |  |  |  |  |  |  |  |  |  |  |  |  |
|  |  |  |  |  |  |  |  |  |  |  |  |  |  |
|  |  |  |  |  |  |  |  |  |  |  |  |  |  |
|  |  |  |  |  |  |  |  |  |  |  |  |  |  |
|  |  |  |  |  |  |  |  |  |  |  |  |  |  |
|  |  |  |  |  |  |  |  |  |  |  |  |  |  |
|  |  |  |  |  |  |  |  |  |  |  |  |  |  |
|  |  |  |  |  |  |  |  |  |  |  |  |  |  |
|  |  |  |  |  |  |  |  |  |  |  |  |  |  |
|  |  |  |  |  |  |  |  |  |  |  |  |  |  |
|  |  |  |  |  |  |  |  |  |  |  |  |  |  |

## REMARKS & HAPPENINGS

# SHIP'S LOG

DATE_____CREW_____GUESTS_____PAGE_____

DEPARTURE TIME_____DESTINATION_____ARRIVAL TIME_____

| TIME | ENGINE HOURS | PILOT INT'L | POSITION | HEADING | RPM | SPEED | ENGINE TEMP | AMPS | OIL | FUEL TANK | BARO | TEMP | WIND |
|------|------|------|------|------|------|------|------|------|------|------|------|------|------|
|  |  |  |  |  |  |  |  |  |  |  |  |  |  |
|  |  |  |  |  |  |  |  |  |  |  |  |  |  |
|  |  |  |  |  |  |  |  |  |  |  |  |  |  |
|  |  |  |  |  |  |  |  |  |  |  |  |  |  |
|  |  |  |  |  |  |  |  |  |  |  |  |  |  |
|  |  |  |  |  |  |  |  |  |  |  |  |  |  |
|  |  |  |  |  |  |  |  |  |  |  |  |  |  |
|  |  |  |  |  |  |  |  |  |  |  |  |  |  |
|  |  |  |  |  |  |  |  |  |  |  |  |  |  |
|  |  |  |  |  |  |  |  |  |  |  |  |  |  |
|  |  |  |  |  |  |  |  |  |  |  |  |  |  |
|  |  |  |  |  |  |  |  |  |  |  |  |  |  |
|  |  |  |  |  |  |  |  |  |  |  |  |  |  |
|  |  |  |  |  |  |  |  |  |  |  |  |  |  |
|  |  |  |  |  |  |  |  |  |  |  |  |  |  |
|  |  |  |  |  |  |  |  |  |  |  |  |  |  |
|  |  |  |  |  |  |  |  |  |  |  |  |  |  |

## REMARKS & HAPPENINGS

# SHIP'S LOG

DATE_____CREW_____GUESTS_____PAGE_____

DEPARTURE TIME_____DESTINATION_____ARRIVAL TIME_____

| TIME | ENGINE HOURS | PILOT INT'L | POSITION | HEADING | RPM | SPEED | ENGINE TEMP | AMPS | OIL | FUEL TANK | BARO | TEMP | WIND |
|------|------|------|------|------|------|------|------|------|------|------|------|------|------|
| | | | | | | | | | | | | | |
| | | | | | | | | | | | | | |
| | | | | | | | | | | | | | |
| | | | | | | | | | | | | | |
| | | | | | | | | | | | | | |
| | | | | | | | | | | | | | |
| | | | | | | | | | | | | | |
| | | | | | | | | | | | | | |
| | | | | | | | | | | | | | |
| | | | | | | | | | | | | | |
| | | | | | | | | | | | | | |
| | | | | | | | | | | | | | |
| | | | | | | | | | | | | | |
| | | | | | | | | | | | | | |
| | | | | | | | | | | | | | |

## REMARKS & HAPPENINGS

# SHIP'S LOG

DATE_____CREW_____GUESTS_____PAGE_____

DEPARTURE TIME_____DESTINATION_____ARRIVAL TIME_____

| TIME | ENGINE HOURS | PILOT INT'L | POSITION | HEADING | RPM | SPEED | ENGINE TEMP | AMPS | OIL | FUEL TANK | BARO | TEMP | WIND |
|------|--------------|-------------|----------|---------|-----|-------|-------------|------|-----|-----------|------|------|------|
|      |              |             |          |         |     |       |             |      |     |           |      |      |      |
|      |              |             |          |         |     |       |             |      |     |           |      |      |      |
|      |              |             |          |         |     |       |             |      |     |           |      |      |      |
|      |              |             |          |         |     |       |             |      |     |           |      |      |      |
|      |              |             |          |         |     |       |             |      |     |           |      |      |      |
|      |              |             |          |         |     |       |             |      |     |           |      |      |      |
|      |              |             |          |         |     |       |             |      |     |           |      |      |      |
|      |              |             |          |         |     |       |             |      |     |           |      |      |      |
|      |              |             |          |         |     |       |             |      |     |           |      |      |      |
|      |              |             |          |         |     |       |             |      |     |           |      |      |      |
|      |              |             |          |         |     |       |             |      |     |           |      |      |      |
|      |              |             |          |         |     |       |             |      |     |           |      |      |      |
|      |              |             |          |         |     |       |             |      |     |           |      |      |      |

## REMARKS & HAPPENINGS

# SHIP'S LOG

DATE_____CREW_____GUESTS_____PAGE_____

DEPARTURE TIME_____DESTINATION_____ARRIVAL TIME_____

| TIME | ENGINE HOURS | PILOT INT'L | POSITION | HEADING | RPM | SPEED | ENGINE TEMP | AMPS | OIL | FUEL TANK | BARO | TEMP | WIND |
|------|------|------|------|------|------|------|------|------|------|------|------|------|------|
| | | | | | | | | | | | | | |
| | | | | | | | | | | | | | |
| | | | | | | | | | | | | | |
| | | | | | | | | | | | | | |
| | | | | | | | | | | | | | |
| | | | | | | | | | | | | | |
| | | | | | | | | | | | | | |
| | | | | | | | | | | | | | |
| | | | | | | | | | | | | | |
| | | | | | | | | | | | | | |
| | | | | | | | | | | | | | |
| | | | | | | | | | | | | | |
| | | | | | | | | | | | | | |
| | | | | | | | | | | | | | |
| | | | | | | | | | | | | | |

## REMARKS & HAPPENINGS

# SHIP'S LOG

DATE_____CREW_____GUESTS_____PAGE_____

DEPARTURE TIME_____DESTINATION_____ARRIVAL TIME_____

| TIME | ENGINE HOURS | PILOT INT'L | POSITION | HEADING | RPM | SPEED | ENGINE TEMP | AMPS | OIL | FUEL TANK | BARO | TEMP | WIND |
|------|------|------|------|------|------|------|------|------|------|------|------|------|------|
|  |  |  |  |  |  |  |  |  |  |  |  |  |  |
|  |  |  |  |  |  |  |  |  |  |  |  |  |  |
|  |  |  |  |  |  |  |  |  |  |  |  |  |  |
|  |  |  |  |  |  |  |  |  |  |  |  |  |  |
|  |  |  |  |  |  |  |  |  |  |  |  |  |  |
|  |  |  |  |  |  |  |  |  |  |  |  |  |  |
|  |  |  |  |  |  |  |  |  |  |  |  |  |  |
|  |  |  |  |  |  |  |  |  |  |  |  |  |  |
|  |  |  |  |  |  |  |  |  |  |  |  |  |  |
|  |  |  |  |  |  |  |  |  |  |  |  |  |  |
|  |  |  |  |  |  |  |  |  |  |  |  |  |  |
|  |  |  |  |  |  |  |  |  |  |  |  |  |  |

## REMARKS & HAPPENINGS

# SHIP'S LOG

DATE_____CREW_____GUESTS_____PAGE_____

DEPARTURE TIME_____DESTINATION_____ARRIVAL TIME_____

| TIME | ENGINE HOURS | PILOT INT'L | POSITION | HEADING | RPM | SPEED | ENGINE TEMP | AMPS | OIL | FUEL TANK | BARO | TEMP | WIND |
|------|------|------|------|------|------|------|------|------|------|------|------|------|------|
|  |  |  |  |  |  |  |  |  |  |  |  |  |  |
|  |  |  |  |  |  |  |  |  |  |  |  |  |  |
|  |  |  |  |  |  |  |  |  |  |  |  |  |  |
|  |  |  |  |  |  |  |  |  |  |  |  |  |  |
|  |  |  |  |  |  |  |  |  |  |  |  |  |  |
|  |  |  |  |  |  |  |  |  |  |  |  |  |  |
|  |  |  |  |  |  |  |  |  |  |  |  |  |  |
|  |  |  |  |  |  |  |  |  |  |  |  |  |  |
|  |  |  |  |  |  |  |  |  |  |  |  |  |  |
|  |  |  |  |  |  |  |  |  |  |  |  |  |  |
|  |  |  |  |  |  |  |  |  |  |  |  |  |  |
|  |  |  |  |  |  |  |  |  |  |  |  |  |  |
|  |  |  |  |  |  |  |  |  |  |  |  |  |  |

## REMARKS & HAPPENINGS

# SHIP'S LOG

DATE_____CREW_____GUESTS_____PAGE_____

DEPARTURE TIME_____DESTINATION_____ARRIVAL TIME_____

| TIME | ENGINE HOURS | PILOT INT'L | POSITION | HEADING | RPM | SPEED | ENGINE TEMP | AMPS | OIL | FUEL TANK | BARO | TEMP | WIND |
|------|------|------|------|------|------|------|------|------|------|------|------|------|------|
|  |  |  |  |  |  |  |  |  |  |  |  |  |  |
|  |  |  |  |  |  |  |  |  |  |  |  |  |  |
|  |  |  |  |  |  |  |  |  |  |  |  |  |  |
|  |  |  |  |  |  |  |  |  |  |  |  |  |  |
|  |  |  |  |  |  |  |  |  |  |  |  |  |  |
|  |  |  |  |  |  |  |  |  |  |  |  |  |  |
|  |  |  |  |  |  |  |  |  |  |  |  |  |  |
|  |  |  |  |  |  |  |  |  |  |  |  |  |  |
|  |  |  |  |  |  |  |  |  |  |  |  |  |  |
|  |  |  |  |  |  |  |  |  |  |  |  |  |  |
|  |  |  |  |  |  |  |  |  |  |  |  |  |  |
|  |  |  |  |  |  |  |  |  |  |  |  |  |  |
|  |  |  |  |  |  |  |  |  |  |  |  |  |  |

## REMARKS & HAPPENINGS

# SHIP'S LOG

DATE_____CREW_____GUESTS_____PAGE_____

DEPARTURE TIME_____DESTINATION_____ARRIVAL TIME_____

| TIME | ENGINE HOURS | PILOT INT'L | POSITION | HEADING | RPM | SPEED | ENGINE TEMP | AMPS | OIL | FUEL TANK | BARO | TEMP | WIND |
|------|------|------|------|------|------|------|------|------|------|------|------|------|------|
|  |  |  |  |  |  |  |  |  |  |  |  |  |  |
|  |  |  |  |  |  |  |  |  |  |  |  |  |  |
|  |  |  |  |  |  |  |  |  |  |  |  |  |  |
|  |  |  |  |  |  |  |  |  |  |  |  |  |  |
|  |  |  |  |  |  |  |  |  |  |  |  |  |  |
|  |  |  |  |  |  |  |  |  |  |  |  |  |  |
|  |  |  |  |  |  |  |  |  |  |  |  |  |  |
|  |  |  |  |  |  |  |  |  |  |  |  |  |  |
|  |  |  |  |  |  |  |  |  |  |  |  |  |  |
|  |  |  |  |  |  |  |  |  |  |  |  |  |  |
|  |  |  |  |  |  |  |  |  |  |  |  |  |  |
|  |  |  |  |  |  |  |  |  |  |  |  |  |  |
|  |  |  |  |  |  |  |  |  |  |  |  |  |  |

## REMARKS & HAPPENINGS

# SHIP'S LOG

DATE_____CREW_____GUESTS_____PAGE_____

DEPARTURE TIME_____DESTINATION_____ARRIVAL TIME_____

| TIME | ENGINE HOURS | PILOT INT'L | POSITION | HEADING | RPM | SPEED | ENGINE TEMP | AMPS | OIL | FUEL TANK | BARO | TEMP | WIND |
|------|------|------|------|------|------|------|------|------|------|------|------|------|------|
| | | | | | | | | | | | | | |
| | | | | | | | | | | | | | |
| | | | | | | | | | | | | | |
| | | | | | | | | | | | | | |
| | | | | | | | | | | | | | |
| | | | | | | | | | | | | | |
| | | | | | | | | | | | | | |
| | | | | | | | | | | | | | |
| | | | | | | | | | | | | | |
| | | | | | | | | | | | | | |
| | | | | | | | | | | | | | |
| | | | | | | | | | | | | | |
| | | | | | | | | | | | | | |
| | | | | | | | | | | | | | |
| | | | | | | | | | | | | | |
| | | | | | | | | | | | | | |

## REMARKS & HAPPENINGS

# SHIP'S LOG

DATE_____CREW_____GUESTS_____PAGE_____

DEPARTURE TIME_____DESTINATION_____ARRIVAL TIME_____

| TIME | ENGINE HOURS | PILOT INT'L | POSITION | HEADING | RPM | SPEED | ENGINE TEMP | AMPS | OIL | FUEL TANK | BARO | TEMP | WIND |
|------|------|------|------|------|------|------|------|------|------|------|------|------|------|
| | | | | | | | | | | | | | |
| | | | | | | | | | | | | | |
| | | | | | | | | | | | | | |
| | | | | | | | | | | | | | |
| | | | | | | | | | | | | | |
| | | | | | | | | | | | | | |
| | | | | | | | | | | | | | |
| | | | | | | | | | | | | | |
| | | | | | | | | | | | | | |
| | | | | | | | | | | | | | |
| | | | | | | | | | | | | | |
| | | | | | | | | | | | | | |
| | | | | | | | | | | | | | |
| | | | | | | | | | | | | | |
| | | | | | | | | | | | | | |

## REMARKS & HAPPENINGS

# SHIP'S LOG

DATE_____CREW_____GUESTS_____PAGE_____

DEPARTURE TIME_____DESTINATION_____ARRIVAL TIME_____

| TIME | ENGINE HOURS | PILOT INT'L | POSITION | HEADING | RPM | SPEED | ENGINE TEMP | AMPS | OIL | FUEL TANK | BARO | TEMP | WIND |
|------|------|------|------|------|------|------|------|------|------|------|------|------|------|
| | | | | | | | | | | | | | |
| | | | | | | | | | | | | | |
| | | | | | | | | | | | | | |
| | | | | | | | | | | | | | |
| | | | | | | | | | | | | | |
| | | | | | | | | | | | | | |
| | | | | | | | | | | | | | |
| | | | | | | | | | | | | | |
| | | | | | | | | | | | | | |
| | | | | | | | | | | | | | |
| | | | | | | | | | | | | | |
| | | | | | | | | | | | | | |

## REMARKS & HAPPENINGS

# SHIP'S LOG

DATE_____CREW_____GUESTS_____PAGE_____

DEPARTURE TIME_____DESTINATION_____ARRIVAL TIME_____

| TIME | ENGINE HOURS | PILOT INT'L | POSITION | HEADING | RPM | SPEED | ENGINE TEMP | AMPS | OIL | FUEL TANK | BARO | TEMP | WIND |
|------|------|------|------|------|------|------|------|------|------|------|------|------|------|
|  |  |  |  |  |  |  |  |  |  |  |  |  |  |
|  |  |  |  |  |  |  |  |  |  |  |  |  |  |
|  |  |  |  |  |  |  |  |  |  |  |  |  |  |
|  |  |  |  |  |  |  |  |  |  |  |  |  |  |
|  |  |  |  |  |  |  |  |  |  |  |  |  |  |
|  |  |  |  |  |  |  |  |  |  |  |  |  |  |
|  |  |  |  |  |  |  |  |  |  |  |  |  |  |
|  |  |  |  |  |  |  |  |  |  |  |  |  |  |
|  |  |  |  |  |  |  |  |  |  |  |  |  |  |
|  |  |  |  |  |  |  |  |  |  |  |  |  |  |
|  |  |  |  |  |  |  |  |  |  |  |  |  |  |
|  |  |  |  |  |  |  |  |  |  |  |  |  |  |
|  |  |  |  |  |  |  |  |  |  |  |  |  |  |
|  |  |  |  |  |  |  |  |  |  |  |  |  |  |
|  |  |  |  |  |  |  |  |  |  |  |  |  |  |
|  |  |  |  |  |  |  |  |  |  |  |  |  |  |
|  |  |  |  |  |  |  |  |  |  |  |  |  |  |

## REMARKS & HAPPENINGS

# SHIP'S LOG

DATE_____CREW_____GUESTS_____PAGE_____

DEPARTURE TIME_____DESTINATION_____ARRIVAL TIME_____

| TIME | ENGINE HOURS | PILOT INT'L | POSITION | HEADING | RPM | SPEED | ENGINE TEMP | AMPS | OIL | FUEL TANK | BARO | TEMP | WIND |
|------|------|------|------|------|------|------|------|------|------|------|------|------|------|
|      |      |      |      |      |      |      |      |      |      |      |      |      |      |
|      |      |      |      |      |      |      |      |      |      |      |      |      |      |
|      |      |      |      |      |      |      |      |      |      |      |      |      |      |
|      |      |      |      |      |      |      |      |      |      |      |      |      |      |
|      |      |      |      |      |      |      |      |      |      |      |      |      |      |
|      |      |      |      |      |      |      |      |      |      |      |      |      |      |
|      |      |      |      |      |      |      |      |      |      |      |      |      |      |
|      |      |      |      |      |      |      |      |      |      |      |      |      |      |
|      |      |      |      |      |      |      |      |      |      |      |      |      |      |
|      |      |      |      |      |      |      |      |      |      |      |      |      |      |
|      |      |      |      |      |      |      |      |      |      |      |      |      |      |
|      |      |      |      |      |      |      |      |      |      |      |      |      |      |

## REMARKS & HAPPENINGS

# SHIP'S LOG

DATE_____CREW_____GUESTS_____PAGE_____

DEPARTURE TIME_____DESTINATION_____ARRIVAL TIME_____

| TIME | ENGINE HOURS | PILOT INT'L | POSITION | HEADING | RPM | SPEED | ENGINE TEMP | AMPS | OIL | FUEL TANK | BARO | TEMP | WIND |
|------|------|------|------|------|------|------|------|------|------|------|------|------|------|
| | | | | | | | | | | | | | |
| | | | | | | | | | | | | | |
| | | | | | | | | | | | | | |
| | | | | | | | | | | | | | |
| | | | | | | | | | | | | | |
| | | | | | | | | | | | | | |
| | | | | | | | | | | | | | |
| | | | | | | | | | | | | | |
| | | | | | | | | | | | | | |
| | | | | | | | | | | | | | |
| | | | | | | | | | | | | | |
| | | | | | | | | | | | | | |
| | | | | | | | | | | | | | |
| | | | | | | | | | | | | | |
| | | | | | | | | | | | | | |
| | | | | | | | | | | | | | |
| | | | | | | | | | | | | | |

## REMARKS & HAPPENINGS

# SHIP'S LOG

DATE_____CREW_____GUESTS_____PAGE_____

DEPARTURE TIME_____DESTINATION_____ARRIVAL TIME_____

| TIME | ENGINE HOURS | PILOT INT'L | POSITION | HEADING | RPM | SPEED | ENGINE TEMP | AMPS | OIL | FUEL TANK | BARO | TEMP | WIND |
|------|------|------|------|------|------|------|------|------|------|------|------|------|------|
|  |  |  |  |  |  |  |  |  |  |  |  |  |  |
|  |  |  |  |  |  |  |  |  |  |  |  |  |  |
|  |  |  |  |  |  |  |  |  |  |  |  |  |  |
|  |  |  |  |  |  |  |  |  |  |  |  |  |  |
|  |  |  |  |  |  |  |  |  |  |  |  |  |  |
|  |  |  |  |  |  |  |  |  |  |  |  |  |  |
|  |  |  |  |  |  |  |  |  |  |  |  |  |  |
|  |  |  |  |  |  |  |  |  |  |  |  |  |  |
|  |  |  |  |  |  |  |  |  |  |  |  |  |  |
|  |  |  |  |  |  |  |  |  |  |  |  |  |  |
|  |  |  |  |  |  |  |  |  |  |  |  |  |  |
|  |  |  |  |  |  |  |  |  |  |  |  |  |  |
|  |  |  |  |  |  |  |  |  |  |  |  |  |  |
|  |  |  |  |  |  |  |  |  |  |  |  |  |  |

## REMARKS & HAPPENINGS

# SHIP'S LOG

DATE_____CREW_____GUESTS_____PAGE_____

DEPARTURE TIME_____DESTINATION_____ARRIVAL TIME_____

| TIME | ENGINE HOURS | PILOT INT'L | POSITION | HEADING | RPM | SPEED | ENGINE TEMP | AMPS | OIL | FUEL TANK | BARO | TEMP | WIND |
|------|------|------|------|------|------|------|------|------|------|------|------|------|------|
| | | | | | | | | | | | | | |
| | | | | | | | | | | | | | |
| | | | | | | | | | | | | | |
| | | | | | | | | | | | | | |
| | | | | | | | | | | | | | |
| | | | | | | | | | | | | | |
| | | | | | | | | | | | | | |
| | | | | | | | | | | | | | |
| | | | | | | | | | | | | | |
| | | | | | | | | | | | | | |
| | | | | | | | | | | | | | |
| | | | | | | | | | | | | | |
| | | | | | | | | | | | | | |
| | | | | | | | | | | | | | |

## REMARKS & HAPPENINGS

# SHIP'S LOG

DATE_____CREW_____GUESTS_____PAGE_____

DEPARTURE TIME_____DESTINATION_____ARRIVAL TIME_____

| TIME | ENGINE HOURS | PILOT INT'L | POSITION | HEADING | RPM | SPEED | ENGINE TEMP | AMPS | OIL | FUEL TANK | BARO | TEMP | WIND |
|------|------|------|------|------|------|------|------|------|------|------|------|------|------|
|  |  |  |  |  |  |  |  |  |  |  |  |  |  |
|  |  |  |  |  |  |  |  |  |  |  |  |  |  |
|  |  |  |  |  |  |  |  |  |  |  |  |  |  |
|  |  |  |  |  |  |  |  |  |  |  |  |  |  |
|  |  |  |  |  |  |  |  |  |  |  |  |  |  |
|  |  |  |  |  |  |  |  |  |  |  |  |  |  |
|  |  |  |  |  |  |  |  |  |  |  |  |  |  |
|  |  |  |  |  |  |  |  |  |  |  |  |  |  |
|  |  |  |  |  |  |  |  |  |  |  |  |  |  |
|  |  |  |  |  |  |  |  |  |  |  |  |  |  |
|  |  |  |  |  |  |  |  |  |  |  |  |  |  |
|  |  |  |  |  |  |  |  |  |  |  |  |  |  |
|  |  |  |  |  |  |  |  |  |  |  |  |  |  |
|  |  |  |  |  |  |  |  |  |  |  |  |  |  |

## REMARKS & HAPPENINGS

# SHIP'S LOG

DATE_____CREW_____GUESTS_____PAGE_____

DEPARTURE TIME_____DESTINATION_____ARRIVAL TIME_____

| TIME | ENGINE HOURS | PILOT INT'L | POSITION | HEADING | RPM | SPEED | ENGINE TEMP | AMPS | OIL | FUEL TANK | BARO | TEMP | WIND |
|------|------|------|------|------|------|------|------|------|------|------|------|------|------|
| | | | | | | | | | | | | | |
| | | | | | | | | | | | | | |
| | | | | | | | | | | | | | |
| | | | | | | | | | | | | | |
| | | | | | | | | | | | | | |
| | | | | | | | | | | | | | |
| | | | | | | | | | | | | | |
| | | | | | | | | | | | | | |
| | | | | | | | | | | | | | |
| | | | | | | | | | | | | | |
| | | | | | | | | | | | | | |
| | | | | | | | | | | | | | |
| | | | | | | | | | | | | | |
| | | | | | | | | | | | | | |
| | | | | | | | | | | | | | |
| | | | | | | | | | | | | | |
| | | | | | | | | | | | | | |
| | | | | | | | | | | | | | |

## REMARKS & HAPPENINGS

# SHIP'S LOG

DATE_____CREW_____GUESTS_____PAGE_____

DEPARTURE TIME_____DESTINATION_____ARRIVAL TIME_____

| TIME | ENGINE HOURS | PILOT INT'L | POSITION | HEADING | RPM | SPEED | ENGINE TEMP | AMPS | OIL | FUEL TANK | BARO | TEMP | WIND |
|------|------|------|------|------|------|------|------|------|------|------|------|------|------|
| | | | | | | | | | | | | | |
| | | | | | | | | | | | | | |
| | | | | | | | | | | | | | |
| | | | | | | | | | | | | | |
| | | | | | | | | | | | | | |
| | | | | | | | | | | | | | |
| | | | | | | | | | | | | | |
| | | | | | | | | | | | | | |
| | | | | | | | | | | | | | |
| | | | | | | | | | | | | | |
| | | | | | | | | | | | | | |
| | | | | | | | | | | | | | |
| | | | | | | | | | | | | | |
| | | | | | | | | | | | | | |

REMARKS & HAPPENINGS

# SHIP'S LOG

DATE_____CREW_____GUESTS_____PAGE_____

DEPARTURE TIME_____DESTINATION_____ARRIVAL TIME_____

| TIME | ENGINE HOURS | PILOT INT'L | POSITION | HEADING | RPM | SPEED | ENGINE TEMP | AMPS | OIL | FUEL TANK | BARO | TEMP | WIND |
|------|------|------|------|------|------|------|------|------|------|------|------|------|------|
|  |  |  |  |  |  |  |  |  |  |  |  |  |  |
|  |  |  |  |  |  |  |  |  |  |  |  |  |  |
|  |  |  |  |  |  |  |  |  |  |  |  |  |  |
|  |  |  |  |  |  |  |  |  |  |  |  |  |  |
|  |  |  |  |  |  |  |  |  |  |  |  |  |  |
|  |  |  |  |  |  |  |  |  |  |  |  |  |  |
|  |  |  |  |  |  |  |  |  |  |  |  |  |  |
|  |  |  |  |  |  |  |  |  |  |  |  |  |  |
|  |  |  |  |  |  |  |  |  |  |  |  |  |  |
|  |  |  |  |  |  |  |  |  |  |  |  |  |  |
|  |  |  |  |  |  |  |  |  |  |  |  |  |  |
|  |  |  |  |  |  |  |  |  |  |  |  |  |  |
|  |  |  |  |  |  |  |  |  |  |  |  |  |  |
|  |  |  |  |  |  |  |  |  |  |  |  |  |  |
|  |  |  |  |  |  |  |  |  |  |  |  |  |  |

## REMARKS & HAPPENINGS

# SHIP'S LOG

DATE_____CREW_____GUESTS_____PAGE_____

DEPARTURE TIME_____DESTINATION_____ARRIVAL TIME_____

| TIME | ENGINE HOURS | PILOT INT'L | POSITION | HEADING | RPM | SPEED | ENGINE TEMP | AMPS | OIL | FUEL TANK | BARO | TEMP | WIND |
|------|------|------|------|------|------|------|------|------|------|------|------|------|------|
|  |  |  |  |  |  |  |  |  |  |  |  |  |  |
|  |  |  |  |  |  |  |  |  |  |  |  |  |  |
|  |  |  |  |  |  |  |  |  |  |  |  |  |  |
|  |  |  |  |  |  |  |  |  |  |  |  |  |  |
|  |  |  |  |  |  |  |  |  |  |  |  |  |  |
|  |  |  |  |  |  |  |  |  |  |  |  |  |  |
|  |  |  |  |  |  |  |  |  |  |  |  |  |  |
|  |  |  |  |  |  |  |  |  |  |  |  |  |  |
|  |  |  |  |  |  |  |  |  |  |  |  |  |  |
|  |  |  |  |  |  |  |  |  |  |  |  |  |  |
|  |  |  |  |  |  |  |  |  |  |  |  |  |  |
|  |  |  |  |  |  |  |  |  |  |  |  |  |  |
|  |  |  |  |  |  |  |  |  |  |  |  |  |  |

## REMARKS & HAPPENINGS

# SHIP'S LOG

DATE_____CREW_____GUESTS_____PAGE_____

DEPARTURE TIME_____DESTINATION_____ARRIVAL TIME_____

| TIME | ENGINE HOURS | PILOT INT'L | POSITION | HEADING | RPM | SPEED | ENGINE TEMP | AMPS | OIL | FUEL TANK | BARO | TEMP | WIND |
|------|------|------|------|------|------|------|------|------|------|------|------|------|------|
| | | | | | | | | | | | | | |
| | | | | | | | | | | | | | |
| | | | | | | | | | | | | | |
| | | | | | | | | | | | | | |
| | | | | | | | | | | | | | |
| | | | | | | | | | | | | | |
| | | | | | | | | | | | | | |
| | | | | | | | | | | | | | |
| | | | | | | | | | | | | | |
| | | | | | | | | | | | | | |
| | | | | | | | | | | | | | |
| | | | | | | | | | | | | | |
| | | | | | | | | | | | | | |
| | | | | | | | | | | | | | |
| | | | | | | | | | | | | | |
| | | | | | | | | | | | | | |

## REMARKS & HAPPENINGS

# SHIP'S LOG

DATE_____CREW_____GUESTS_____PAGE_____

DEPARTURE TIME_____DESTINATION_____ARRIVAL TIME_____

| TIME | ENGINE HOURS | PILOT INT'L | POSITION | HEADING | RPM | SPEED | ENGINE TEMP | AMPS | OIL | FUEL TANK | BARO | TEMP | WIND |
|---|---|---|---|---|---|---|---|---|---|---|---|---|---|
|  |  |  |  |  |  |  |  |  |  |  |  |  |  |
|  |  |  |  |  |  |  |  |  |  |  |  |  |  |
|  |  |  |  |  |  |  |  |  |  |  |  |  |  |
|  |  |  |  |  |  |  |  |  |  |  |  |  |  |
|  |  |  |  |  |  |  |  |  |  |  |  |  |  |
|  |  |  |  |  |  |  |  |  |  |  |  |  |  |
|  |  |  |  |  |  |  |  |  |  |  |  |  |  |
|  |  |  |  |  |  |  |  |  |  |  |  |  |  |
|  |  |  |  |  |  |  |  |  |  |  |  |  |  |
|  |  |  |  |  |  |  |  |  |  |  |  |  |  |
|  |  |  |  |  |  |  |  |  |  |  |  |  |  |
|  |  |  |  |  |  |  |  |  |  |  |  |  |  |
|  |  |  |  |  |  |  |  |  |  |  |  |  |  |

## REMARKS & HAPPENINGS

# SHIP'S LOG

DATE_____CREW_____GUESTS_____PAGE_____

DEPARTURE TIME_____DESTINATION_____ARRIVAL TIME_____

| TIME | ENGINE HOURS | PILOT INT'L | POSITION | HEADING | RPM | SPEED | ENGINE TEMP | AMPS | OIL | FUEL TANK | BARO | TEMP | WIND |
|------|------|------|------|------|------|------|------|------|------|------|------|------|------|
|  |  |  |  |  |  |  |  |  |  |  |  |  |  |
|  |  |  |  |  |  |  |  |  |  |  |  |  |  |
|  |  |  |  |  |  |  |  |  |  |  |  |  |  |
|  |  |  |  |  |  |  |  |  |  |  |  |  |  |
|  |  |  |  |  |  |  |  |  |  |  |  |  |  |
|  |  |  |  |  |  |  |  |  |  |  |  |  |  |
|  |  |  |  |  |  |  |  |  |  |  |  |  |  |
|  |  |  |  |  |  |  |  |  |  |  |  |  |  |
|  |  |  |  |  |  |  |  |  |  |  |  |  |  |
|  |  |  |  |  |  |  |  |  |  |  |  |  |  |
|  |  |  |  |  |  |  |  |  |  |  |  |  |  |
|  |  |  |  |  |  |  |  |  |  |  |  |  |  |
|  |  |  |  |  |  |  |  |  |  |  |  |  |  |
|  |  |  |  |  |  |  |  |  |  |  |  |  |  |
|  |  |  |  |  |  |  |  |  |  |  |  |  |  |
|  |  |  |  |  |  |  |  |  |  |  |  |  |  |

## REMARKS & HAPPENINGS

# SHIP'S LOG

DATE_____ CREW_____ GUESTS_____ PAGE_____

DEPARTURE TIME_____ DESTINATION_____ ARRIVAL TIME_____

| TIME | ENGINE HOURS | PILOT INT'L | POSITION | HEADING | RPM | SPEED | ENGINE TEMP | AMPS | OIL | FUEL TANK | BARO | TEMP | WIND |
|------|------|------|------|------|------|------|------|------|------|------|------|------|------|
| | | | | | | | | | | | | | |
| | | | | | | | | | | | | | |
| | | | | | | | | | | | | | |
| | | | | | | | | | | | | | |
| | | | | | | | | | | | | | |
| | | | | | | | | | | | | | |
| | | | | | | | | | | | | | |
| | | | | | | | | | | | | | |
| | | | | | | | | | | | | | |
| | | | | | | | | | | | | | |
| | | | | | | | | | | | | | |
| | | | | | | | | | | | | | |
| | | | | | | | | | | | | | |
| | | | | | | | | | | | | | |

## REMARKS & HAPPENINGS

# SHIP'S LOG

DATE_____CREW_____GUESTS_____PAGE_____

DEPARTURE TIME_____DESTINATION_____ARRIVAL TIME_____

| TIME | ENGINE HOURS | PILOT INT'L | POSITION | HEADING | RPM | SPEED | ENGINE TEMP | AMPS | OIL | FUEL TANK | BARO | TEMP | WIND |
|------|--------------|-------------|----------|---------|-----|-------|-------------|------|-----|-----------|------|------|------|
|      |              |             |          |         |     |       |             |      |     |           |      |      |      |
|      |              |             |          |         |     |       |             |      |     |           |      |      |      |
|      |              |             |          |         |     |       |             |      |     |           |      |      |      |
|      |              |             |          |         |     |       |             |      |     |           |      |      |      |
|      |              |             |          |         |     |       |             |      |     |           |      |      |      |
|      |              |             |          |         |     |       |             |      |     |           |      |      |      |
|      |              |             |          |         |     |       |             |      |     |           |      |      |      |
|      |              |             |          |         |     |       |             |      |     |           |      |      |      |
|      |              |             |          |         |     |       |             |      |     |           |      |      |      |
|      |              |             |          |         |     |       |             |      |     |           |      |      |      |
|      |              |             |          |         |     |       |             |      |     |           |      |      |      |
|      |              |             |          |         |     |       |             |      |     |           |      |      |      |
|      |              |             |          |         |     |       |             |      |     |           |      |      |      |
|      |              |             |          |         |     |       |             |      |     |           |      |      |      |

## REMARKS & HAPPENINGS

# SHIP'S LOG

DATE_____CREW_____GUESTS_____PAGE_____

DEPARTURE TIME_____DESTINATION_____ARRIVAL TIME_____

| TIME | ENGINE HOURS | PILOT INT'L | POSITION | HEADING | RPM | SPEED | ENGINE TEMP | AMPS | OIL | FUEL TANK | BARO | TEMP | WIND |
|------|------|------|------|------|------|------|------|------|------|------|------|------|------|
| | | | | | | | | | | | | | |
| | | | | | | | | | | | | | |
| | | | | | | | | | | | | | |
| | | | | | | | | | | | | | |
| | | | | | | | | | | | | | |
| | | | | | | | | | | | | | |
| | | | | | | | | | | | | | |
| | | | | | | | | | | | | | |
| | | | | | | | | | | | | | |
| | | | | | | | | | | | | | |
| | | | | | | | | | | | | | |
| | | | | | | | | | | | | | |
| | | | | | | | | | | | | | |
| | | | | | | | | | | | | | |
| | | | | | | | | | | | | | |

## REMARKS & HAPPENINGS

# SHIP'S LOG

DATE_____CREW_____GUESTS_____PAGE_____

DEPARTURE TIME_____DESTINATION_____ARRIVAL TIME_____

| TIME | ENGINE HOURS | PILOT INT'L | POSITION | HEADING | RPM | SPEED | ENGINE TEMP | AMPS | OIL | FUEL TANK | BARO | TEMP | WIND |
|------|--------------|-------------|----------|---------|-----|-------|-------------|------|-----|-----------|------|------|------|
|      |              |             |          |         |     |       |             |      |     |           |      |      |      |
|      |              |             |          |         |     |       |             |      |     |           |      |      |      |
|      |              |             |          |         |     |       |             |      |     |           |      |      |      |
|      |              |             |          |         |     |       |             |      |     |           |      |      |      |
|      |              |             |          |         |     |       |             |      |     |           |      |      |      |
|      |              |             |          |         |     |       |             |      |     |           |      |      |      |
|      |              |             |          |         |     |       |             |      |     |           |      |      |      |
|      |              |             |          |         |     |       |             |      |     |           |      |      |      |
|      |              |             |          |         |     |       |             |      |     |           |      |      |      |
|      |              |             |          |         |     |       |             |      |     |           |      |      |      |
|      |              |             |          |         |     |       |             |      |     |           |      |      |      |
|      |              |             |          |         |     |       |             |      |     |           |      |      |      |
|      |              |             |          |         |     |       |             |      |     |           |      |      |      |
|      |              |             |          |         |     |       |             |      |     |           |      |      |      |

## REMARKS & HAPPENINGS

# SHIP'S LOG

DATE_____CREW_____GUESTS_____PAGE_____

DEPARTURE TIME_____DESTINATION_____ARRIVAL TIME_____

| TIME | ENGINE HOURS | PILOT INT'L | POSITION | HEADING | RPM | SPEED | ENGINE TEMP | AMPS | OIL | FUEL TANK | BARO | TEMP | WIND |
|------|------|------|------|------|------|------|------|------|------|------|------|------|------|
|  |  |  |  |  |  |  |  |  |  |  |  |  |  |
|  |  |  |  |  |  |  |  |  |  |  |  |  |  |
|  |  |  |  |  |  |  |  |  |  |  |  |  |  |
|  |  |  |  |  |  |  |  |  |  |  |  |  |  |
|  |  |  |  |  |  |  |  |  |  |  |  |  |  |
|  |  |  |  |  |  |  |  |  |  |  |  |  |  |
|  |  |  |  |  |  |  |  |  |  |  |  |  |  |
|  |  |  |  |  |  |  |  |  |  |  |  |  |  |
|  |  |  |  |  |  |  |  |  |  |  |  |  |  |
|  |  |  |  |  |  |  |  |  |  |  |  |  |  |
|  |  |  |  |  |  |  |  |  |  |  |  |  |  |
|  |  |  |  |  |  |  |  |  |  |  |  |  |  |
|  |  |  |  |  |  |  |  |  |  |  |  |  |  |
|  |  |  |  |  |  |  |  |  |  |  |  |  |  |

## REMARKS & HAPPENINGS

# SHIP'S LOG

DATE_____CREW_____GUESTS_____PAGE_____

DEPARTURE TIME_____DESTINATION_____ARRIVAL TIME_____

| TIME | ENGINE HOURS | PILOT INT'L | POSITION | HEADING | RPM | SPEED | ENGINE TEMP | AMPS | OIL | FUEL TANK | BARO | TEMP | WIND |
|------|------|------|------|------|------|------|------|------|------|------|------|------|------|
| | | | | | | | | | | | | | |
| | | | | | | | | | | | | | |
| | | | | | | | | | | | | | |
| | | | | | | | | | | | | | |
| | | | | | | | | | | | | | |
| | | | | | | | | | | | | | |
| | | | | | | | | | | | | | |
| | | | | | | | | | | | | | |
| | | | | | | | | | | | | | |
| | | | | | | | | | | | | | |
| | | | | | | | | | | | | | |
| | | | | | | | | | | | | | |
| | | | | | | | | | | | | | |
| | | | | | | | | | | | | | |
| | | | | | | | | | | | | | |

## REMARKS & HAPPENINGS

# SHIP'S LOG

DATE_____CREW_____GUESTS_____PAGE_____

DEPARTURE TIME_____DESTINATION_____ARRIVAL TIME_____

| TIME | ENGINE HOURS | PILOT INT'L | POSITION | HEADING | RPM | SPEED | ENGINE TEMP | AMPS | OIL | FUEL TANK | BARO | TEMP | WIND |
|---|---|---|---|---|---|---|---|---|---|---|---|---|---|
|  |  |  |  |  |  |  |  |  |  |  |  |  |  |
|  |  |  |  |  |  |  |  |  |  |  |  |  |  |
|  |  |  |  |  |  |  |  |  |  |  |  |  |  |
|  |  |  |  |  |  |  |  |  |  |  |  |  |  |
|  |  |  |  |  |  |  |  |  |  |  |  |  |  |
|  |  |  |  |  |  |  |  |  |  |  |  |  |  |
|  |  |  |  |  |  |  |  |  |  |  |  |  |  |
|  |  |  |  |  |  |  |  |  |  |  |  |  |  |
|  |  |  |  |  |  |  |  |  |  |  |  |  |  |
|  |  |  |  |  |  |  |  |  |  |  |  |  |  |
|  |  |  |  |  |  |  |  |  |  |  |  |  |  |
|  |  |  |  |  |  |  |  |  |  |  |  |  |  |
|  |  |  |  |  |  |  |  |  |  |  |  |  |  |

## REMARKS & HAPPENINGS

# SHIP'S LOG

DATE_____CREW_____GUESTS_____PAGE_____

DEPARTURE TIME_____DESTINATION_____ARRIVAL TIME_____

| TIME | ENGINE HOURS | PILOT INT'L | POSITION | HEADING | RPM | SPEED | ENGINE TEMP | AMPS | OIL | FUEL TANK | BARO | TEMP | WIND |
|------|------|------|------|------|------|------|------|------|------|------|------|------|------|
|      |      |      |      |      |      |      |      |      |      |      |      |      |      |
|      |      |      |      |      |      |      |      |      |      |      |      |      |      |
|      |      |      |      |      |      |      |      |      |      |      |      |      |      |
|      |      |      |      |      |      |      |      |      |      |      |      |      |      |
|      |      |      |      |      |      |      |      |      |      |      |      |      |      |
|      |      |      |      |      |      |      |      |      |      |      |      |      |      |
|      |      |      |      |      |      |      |      |      |      |      |      |      |      |
|      |      |      |      |      |      |      |      |      |      |      |      |      |      |
|      |      |      |      |      |      |      |      |      |      |      |      |      |      |
|      |      |      |      |      |      |      |      |      |      |      |      |      |      |
|      |      |      |      |      |      |      |      |      |      |      |      |      |      |
|      |      |      |      |      |      |      |      |      |      |      |      |      |      |
|      |      |      |      |      |      |      |      |      |      |      |      |      |      |
|      |      |      |      |      |      |      |      |      |      |      |      |      |      |

## REMARKS & HAPPENINGS

# SHIP'S LOG

DATE_____CREW_____GUESTS_____PAGE_____

DEPARTURE TIME_____DESTINATION_____ARRIVAL TIME_____

| TIME | ENGINE HOURS | PILOT INT'L | POSITION | HEADING | RPM | SPEED | ENGINE TEMP | AMPS | OIL | FUEL TANK | BARO | TEMP | WIND |
|------|------|------|------|------|------|------|------|------|------|------|------|------|------|
|  |  |  |  |  |  |  |  |  |  |  |  |  |  |
|  |  |  |  |  |  |  |  |  |  |  |  |  |  |
|  |  |  |  |  |  |  |  |  |  |  |  |  |  |
|  |  |  |  |  |  |  |  |  |  |  |  |  |  |
|  |  |  |  |  |  |  |  |  |  |  |  |  |  |
|  |  |  |  |  |  |  |  |  |  |  |  |  |  |
|  |  |  |  |  |  |  |  |  |  |  |  |  |  |
|  |  |  |  |  |  |  |  |  |  |  |  |  |  |
|  |  |  |  |  |  |  |  |  |  |  |  |  |  |
|  |  |  |  |  |  |  |  |  |  |  |  |  |  |
|  |  |  |  |  |  |  |  |  |  |  |  |  |  |
|  |  |  |  |  |  |  |  |  |  |  |  |  |  |
|  |  |  |  |  |  |  |  |  |  |  |  |  |  |
|  |  |  |  |  |  |  |  |  |  |  |  |  |  |
|  |  |  |  |  |  |  |  |  |  |  |  |  |  |

## REMARKS & HAPPENINGS

# SHIP'S LOG

DATE_____ CREW_____ GUESTS_____ PAGE_____

DEPARTURE TIME_____ DESTINATION_____ ARRIVAL TIME_____

| TIME | ENGINE HOURS | PILOT INT'L | POSITION | HEADING | RPM | SPEED | ENGINE TEMP | AMPS | OIL | FUEL TANK | BARO | TEMP | WIND |
|------|--------------|-------------|----------|---------|-----|-------|-------------|------|-----|-----------|------|------|------|
|      |              |             |          |         |     |       |             |      |     |           |      |      |      |
|      |              |             |          |         |     |       |             |      |     |           |      |      |      |
|      |              |             |          |         |     |       |             |      |     |           |      |      |      |
|      |              |             |          |         |     |       |             |      |     |           |      |      |      |
|      |              |             |          |         |     |       |             |      |     |           |      |      |      |
|      |              |             |          |         |     |       |             |      |     |           |      |      |      |
|      |              |             |          |         |     |       |             |      |     |           |      |      |      |
|      |              |             |          |         |     |       |             |      |     |           |      |      |      |
|      |              |             |          |         |     |       |             |      |     |           |      |      |      |
|      |              |             |          |         |     |       |             |      |     |           |      |      |      |
|      |              |             |          |         |     |       |             |      |     |           |      |      |      |
|      |              |             |          |         |     |       |             |      |     |           |      |      |      |
|      |              |             |          |         |     |       |             |      |     |           |      |      |      |
|      |              |             |          |         |     |       |             |      |     |           |      |      |      |
|      |              |             |          |         |     |       |             |      |     |           |      |      |      |

## REMARKS & HAPPENINGS

# SHIP'S LOG

DATE_____ CREW_____ GUESTS_____ PAGE_____

DEPARTURE TIME_____ DESTINATION_____ ARRIVAL TIME_____

| TIME | ENGINE HOURS | PILOT INT'L | POSITION | HEADING | RPM | SPEED | ENGINE TEMP | AMPS | OIL | FUEL TANK | BARO | TEMP | WIND |
|------|------|------|------|------|------|------|------|------|------|------|------|------|------|
|      |      |      |      |      |      |      |      |      |      |      |      |      |      |
|      |      |      |      |      |      |      |      |      |      |      |      |      |      |
|      |      |      |      |      |      |      |      |      |      |      |      |      |      |
|      |      |      |      |      |      |      |      |      |      |      |      |      |      |
|      |      |      |      |      |      |      |      |      |      |      |      |      |      |
|      |      |      |      |      |      |      |      |      |      |      |      |      |      |
|      |      |      |      |      |      |      |      |      |      |      |      |      |      |
|      |      |      |      |      |      |      |      |      |      |      |      |      |      |
|      |      |      |      |      |      |      |      |      |      |      |      |      |      |
|      |      |      |      |      |      |      |      |      |      |      |      |      |      |
|      |      |      |      |      |      |      |      |      |      |      |      |      |      |
|      |      |      |      |      |      |      |      |      |      |      |      |      |      |

## REMARKS & HAPPENINGS

# SHIP'S LOG

DATE_____CREW_____GUESTS_____PAGE_____

DEPARTURE TIME_____DESTINATION_____ARRIVAL TIME_____

| TIME | ENGINE HOURS | PILOT INT'L | POSITION | HEADING | RPM | SPEED | ENGINE TEMP | AMPS | OIL | FUEL TANK | BARO | TEMP | WIND |
|------|------|------|------|------|------|------|------|------|------|------|------|------|------|
| | | | | | | | | | | | | | |
| | | | | | | | | | | | | | |
| | | | | | | | | | | | | | |
| | | | | | | | | | | | | | |
| | | | | | | | | | | | | | |
| | | | | | | | | | | | | | |
| | | | | | | | | | | | | | |
| | | | | | | | | | | | | | |
| | | | | | | | | | | | | | |
| | | | | | | | | | | | | | |
| | | | | | | | | | | | | | |
| | | | | | | | | | | | | | |
| | | | | | | | | | | | | | |
| | | | | | | | | | | | | | |
| | | | | | | | | | | | | | |
| | | | | | | | | | | | | | |
| | | | | | | | | | | | | | |
| | | | | | | | | | | | | | |

## REMARKS & HAPPENINGS

# SHIP'S LOG

DATE_____CREW_____GUESTS_____PAGE_____

DEPARTURE TIME_____DESTINATION_____ARRIVAL TIME_____

| TIME | ENGINE HOURS | PILOT INT'L | POSITION | HEADING | RPM | SPEED | ENGINE TEMP | AMPS | OIL | FUEL TANK | BARO | TEMP | WIND |
|------|------|------|------|------|------|------|------|------|------|------|------|------|------|
| | | | | | | | | | | | | | |
| | | | | | | | | | | | | | |
| | | | | | | | | | | | | | |
| | | | | | | | | | | | | | |
| | | | | | | | | | | | | | |
| | | | | | | | | | | | | | |
| | | | | | | | | | | | | | |
| | | | | | | | | | | | | | |
| | | | | | | | | | | | | | |
| | | | | | | | | | | | | | |
| | | | | | | | | | | | | | |
| | | | | | | | | | | | | | |
| | | | | | | | | | | | | | |

## REMARKS & HAPPENINGS

# SHIP'S LOG

DATE_____CREW_____GUESTS_____PAGE_____

DEPARTURE TIME_____DESTINATION_____ARRIVAL TIME_____

| TIME | ENGINE HOURS | PILOT INT'L | POSITION | HEADING | RPM | SPEED | ENGINE TEMP | AMPS | OIL | FUEL TANK | BARO | TEMP | WIND |
|------|------|------|------|------|------|------|------|------|------|------|------|------|------|
| | | | | | | | | | | | | | |
| | | | | | | | | | | | | | |
| | | | | | | | | | | | | | |
| | | | | | | | | | | | | | |
| | | | | | | | | | | | | | |
| | | | | | | | | | | | | | |
| | | | | | | | | | | | | | |
| | | | | | | | | | | | | | |
| | | | | | | | | | | | | | |
| | | | | | | | | | | | | | |
| | | | | | | | | | | | | | |
| | | | | | | | | | | | | | |
| | | | | | | | | | | | | | |
| | | | | | | | | | | | | | |
| | | | | | | | | | | | | | |

## REMARKS & HAPPENINGS

# SHIP'S LOG

DATE_____CREW_____GUESTS_____PAGE_____

DEPARTURE TIME_____DESTINATION_____ARRIVAL TIME_____

| TIME | ENGINE HOURS | PILOT INT'L | POSITION | HEADING | RPM | SPEED | ENGINE TEMP | AMPS | OIL | FUEL TANK | BARO | TEMP | WIND |
|------|------|------|------|------|------|------|------|------|------|------|------|------|------|
|  |  |  |  |  |  |  |  |  |  |  |  |  |  |
|  |  |  |  |  |  |  |  |  |  |  |  |  |  |
|  |  |  |  |  |  |  |  |  |  |  |  |  |  |
|  |  |  |  |  |  |  |  |  |  |  |  |  |  |
|  |  |  |  |  |  |  |  |  |  |  |  |  |  |
|  |  |  |  |  |  |  |  |  |  |  |  |  |  |
|  |  |  |  |  |  |  |  |  |  |  |  |  |  |
|  |  |  |  |  |  |  |  |  |  |  |  |  |  |
|  |  |  |  |  |  |  |  |  |  |  |  |  |  |
|  |  |  |  |  |  |  |  |  |  |  |  |  |  |
|  |  |  |  |  |  |  |  |  |  |  |  |  |  |
|  |  |  |  |  |  |  |  |  |  |  |  |  |  |

## REMARKS & HAPPENINGS

# SHIP'S LOG

DATE_____CREW_____GUESTS_____PAGE_____

DEPARTURE TIME_____DESTINATION_____ARRIVAL TIME_____

| TIME | ENGINE HOURS | PILOT INT'L | POSITION | HEADING | RPM | SPEED | ENGINE TEMP | AMPS | OIL | FUEL TANK | BARO | TEMP | WIND |
|------|------|------|------|------|------|------|------|------|------|------|------|------|------|
|  |  |  |  |  |  |  |  |  |  |  |  |  |  |
|  |  |  |  |  |  |  |  |  |  |  |  |  |  |
|  |  |  |  |  |  |  |  |  |  |  |  |  |  |
|  |  |  |  |  |  |  |  |  |  |  |  |  |  |
|  |  |  |  |  |  |  |  |  |  |  |  |  |  |
|  |  |  |  |  |  |  |  |  |  |  |  |  |  |
|  |  |  |  |  |  |  |  |  |  |  |  |  |  |
|  |  |  |  |  |  |  |  |  |  |  |  |  |  |
|  |  |  |  |  |  |  |  |  |  |  |  |  |  |
|  |  |  |  |  |  |  |  |  |  |  |  |  |  |
|  |  |  |  |  |  |  |  |  |  |  |  |  |  |
|  |  |  |  |  |  |  |  |  |  |  |  |  |  |
|  |  |  |  |  |  |  |  |  |  |  |  |  |  |
|  |  |  |  |  |  |  |  |  |  |  |  |  |  |

## REMARKS & HAPPENINGS

# SHIP'S LOG

DATE_____CREW_____GUESTS_____PAGE_____

DEPARTURE TIME_____DESTINATION_____ARRIVAL TIME_____

| TIME | ENGINE HOURS | PILOT INT'L | POSITION | HEADING | RPM | SPEED | ENGINE TEMP | AMPS | OIL | FUEL TANK | BARO | TEMP | WIND |
|------|------|------|------|------|------|------|------|------|------|------|------|------|------|
|  |  |  |  |  |  |  |  |  |  |  |  |  |  |
|  |  |  |  |  |  |  |  |  |  |  |  |  |  |
|  |  |  |  |  |  |  |  |  |  |  |  |  |  |
|  |  |  |  |  |  |  |  |  |  |  |  |  |  |
|  |  |  |  |  |  |  |  |  |  |  |  |  |  |
|  |  |  |  |  |  |  |  |  |  |  |  |  |  |
|  |  |  |  |  |  |  |  |  |  |  |  |  |  |
|  |  |  |  |  |  |  |  |  |  |  |  |  |  |
|  |  |  |  |  |  |  |  |  |  |  |  |  |  |
|  |  |  |  |  |  |  |  |  |  |  |  |  |  |
|  |  |  |  |  |  |  |  |  |  |  |  |  |  |
|  |  |  |  |  |  |  |  |  |  |  |  |  |  |

## REMARKS & HAPPENINGS

# SHIP'S LOG

DATE_____CREW_____GUESTS_____PAGE_____

DEPARTURE TIME_____DESTINATION_____ARRIVAL TIME_____

| TIME | ENGINE HOURS | PILOT INT'L | POSITION | HEADING | RPM | SPEED | ENGINE TEMP | AMPS | OIL | FUEL TANK | BARO | TEMP | WIND |
|------|------|------|------|------|------|------|------|------|------|------|------|------|------|
|  |  |  |  |  |  |  |  |  |  |  |  |  |  |
|  |  |  |  |  |  |  |  |  |  |  |  |  |  |
|  |  |  |  |  |  |  |  |  |  |  |  |  |  |
|  |  |  |  |  |  |  |  |  |  |  |  |  |  |
|  |  |  |  |  |  |  |  |  |  |  |  |  |  |
|  |  |  |  |  |  |  |  |  |  |  |  |  |  |
|  |  |  |  |  |  |  |  |  |  |  |  |  |  |
|  |  |  |  |  |  |  |  |  |  |  |  |  |  |
|  |  |  |  |  |  |  |  |  |  |  |  |  |  |
|  |  |  |  |  |  |  |  |  |  |  |  |  |  |
|  |  |  |  |  |  |  |  |  |  |  |  |  |  |
|  |  |  |  |  |  |  |  |  |  |  |  |  |  |
|  |  |  |  |  |  |  |  |  |  |  |  |  |  |
|  |  |  |  |  |  |  |  |  |  |  |  |  |  |
|  |  |  |  |  |  |  |  |  |  |  |  |  |  |

## REMARKS & HAPPENINGS

# SHIP'S LOG

DATE_____CREW_____GUESTS_____PAGE_____

DEPARTURE TIME_____DESTINATION_____ARRIVAL TIME_____

| TIME | ENGINE HOURS | PILOT INT'L | POSITION | HEADING | RPM | SPEED | ENGINE TEMP | AMPS | OIL | FUEL TANK | BARO | TEMP | WIND |
|------|------|------|------|------|------|------|------|------|------|------|------|------|------|
| | | | | | | | | | | | | | |
| | | | | | | | | | | | | | |
| | | | | | | | | | | | | | |
| | | | | | | | | | | | | | |
| | | | | | | | | | | | | | |
| | | | | | | | | | | | | | |
| | | | | | | | | | | | | | |
| | | | | | | | | | | | | | |
| | | | | | | | | | | | | | |
| | | | | | | | | | | | | | |
| | | | | | | | | | | | | | |
| | | | | | | | | | | | | | |
| | | | | | | | | | | | | | |
| | | | | | | | | | | | | | |
| | | | | | | | | | | | | | |
| | | | | | | | | | | | | | |

## REMARKS & HAPPENINGS

# SHIP'S LOG

DATE_____CREW_____GUESTS_____PAGE_____

DEPARTURE TIME_____DESTINATION_____ARRIVAL TIME_____

| TIME | ENGINE HOURS | PILOT INT'L | POSITION | HEADING | RPM | SPEED | ENGINE TEMP | AMPS | OIL | FUEL TANK | BARO | TEMP | WIND |
|------|------|------|------|------|------|------|------|------|------|------|------|------|------|
| | | | | | | | | | | | | | |
| | | | | | | | | | | | | | |
| | | | | | | | | | | | | | |
| | | | | | | | | | | | | | |
| | | | | | | | | | | | | | |
| | | | | | | | | | | | | | |
| | | | | | | | | | | | | | |
| | | | | | | | | | | | | | |
| | | | | | | | | | | | | | |
| | | | | | | | | | | | | | |
| | | | | | | | | | | | | | |
| | | | | | | | | | | | | | |
| | | | | | | | | | | | | | |
| | | | | | | | | | | | | | |

## REMARKS & HAPPENINGS

# SHIP'S LOG

DATE_____CREW_____GUESTS_____PAGE_____

DEPARTURE TIME_____DESTINATION_____ARRIVAL TIME_____

| TIME | ENGINE HOURS | PILOT INT'L | POSITION | HEADING | RPM | SPEED | ENGINE TEMP | AMPS | OIL | FUEL TANK | BARO | TEMP | WIND |
|------|------|------|------|------|------|------|------|------|------|------|------|------|------|
|  |  |  |  |  |  |  |  |  |  |  |  |  |  |
|  |  |  |  |  |  |  |  |  |  |  |  |  |  |
|  |  |  |  |  |  |  |  |  |  |  |  |  |  |
|  |  |  |  |  |  |  |  |  |  |  |  |  |  |
|  |  |  |  |  |  |  |  |  |  |  |  |  |  |
|  |  |  |  |  |  |  |  |  |  |  |  |  |  |
|  |  |  |  |  |  |  |  |  |  |  |  |  |  |
|  |  |  |  |  |  |  |  |  |  |  |  |  |  |
|  |  |  |  |  |  |  |  |  |  |  |  |  |  |
|  |  |  |  |  |  |  |  |  |  |  |  |  |  |
|  |  |  |  |  |  |  |  |  |  |  |  |  |  |
|  |  |  |  |  |  |  |  |  |  |  |  |  |  |

## REMARKS & HAPPENINGS

# SHIP'S LOG

DATE_____CREW_____GUESTS_____PAGE_____

DEPARTURE TIME_____DESTINATION_____ARRIVAL TIME_____

| TIME | ENGINE HOURS | PILOT INT'L | POSITION | HEADING | RPM | SPEED | ENGINE TEMP | AMPS | OIL | FUEL TANK | BARO | TEMP | WIND |
|------|------|------|------|------|------|------|------|------|------|------|------|------|------|
|  |  |  |  |  |  |  |  |  |  |  |  |  |  |
|  |  |  |  |  |  |  |  |  |  |  |  |  |  |
|  |  |  |  |  |  |  |  |  |  |  |  |  |  |
|  |  |  |  |  |  |  |  |  |  |  |  |  |  |
|  |  |  |  |  |  |  |  |  |  |  |  |  |  |
|  |  |  |  |  |  |  |  |  |  |  |  |  |  |
|  |  |  |  |  |  |  |  |  |  |  |  |  |  |
|  |  |  |  |  |  |  |  |  |  |  |  |  |  |
|  |  |  |  |  |  |  |  |  |  |  |  |  |  |
|  |  |  |  |  |  |  |  |  |  |  |  |  |  |
|  |  |  |  |  |  |  |  |  |  |  |  |  |  |
|  |  |  |  |  |  |  |  |  |  |  |  |  |  |
|  |  |  |  |  |  |  |  |  |  |  |  |  |  |
|  |  |  |  |  |  |  |  |  |  |  |  |  |  |
|  |  |  |  |  |  |  |  |  |  |  |  |  |  |
|  |  |  |  |  |  |  |  |  |  |  |  |  |  |
|  |  |  |  |  |  |  |  |  |  |  |  |  |  |

## REMARKS & HAPPENINGS

# SHIP'S LOG

DATE_____CREW_____GUESTS_____PAGE_____

DEPARTURE TIME_____DESTINATION_____ARRIVAL TIME_____

| TIME | ENGINE HOURS | PILOT INT'L | POSITION | HEADING | RPM | SPEED | ENGINE TEMP | AMPS | OIL | FUEL TANK | BARO | TEMP | WIND |
|------|------|------|------|------|------|------|------|------|------|------|------|------|------|
|  |  |  |  |  |  |  |  |  |  |  |  |  |  |
|  |  |  |  |  |  |  |  |  |  |  |  |  |  |
|  |  |  |  |  |  |  |  |  |  |  |  |  |  |
|  |  |  |  |  |  |  |  |  |  |  |  |  |  |
|  |  |  |  |  |  |  |  |  |  |  |  |  |  |
|  |  |  |  |  |  |  |  |  |  |  |  |  |  |
|  |  |  |  |  |  |  |  |  |  |  |  |  |  |
|  |  |  |  |  |  |  |  |  |  |  |  |  |  |
|  |  |  |  |  |  |  |  |  |  |  |  |  |  |
|  |  |  |  |  |  |  |  |  |  |  |  |  |  |
|  |  |  |  |  |  |  |  |  |  |  |  |  |  |
|  |  |  |  |  |  |  |  |  |  |  |  |  |  |
|  |  |  |  |  |  |  |  |  |  |  |  |  |  |
|  |  |  |  |  |  |  |  |  |  |  |  |  |  |
|  |  |  |  |  |  |  |  |  |  |  |  |  |  |

## REMARKS & HAPPENINGS

# SHIP'S LOG

DATE_____CREW_____GUESTS_____PAGE_____

DEPARTURE TIME_____DESTINATION_____ARRIVAL TIME_____

| TIME | ENGINE HOURS | PILOT INT'L | POSITION | HEADING | RPM | SPEED | ENGINE TEMP | AMPS | OIL | FUEL TANK | BARO | TEMP | WIND |
|------|------|------|------|------|------|------|------|------|------|------|------|------|------|
|  |  |  |  |  |  |  |  |  |  |  |  |  |  |
|  |  |  |  |  |  |  |  |  |  |  |  |  |  |
|  |  |  |  |  |  |  |  |  |  |  |  |  |  |
|  |  |  |  |  |  |  |  |  |  |  |  |  |  |
|  |  |  |  |  |  |  |  |  |  |  |  |  |  |
|  |  |  |  |  |  |  |  |  |  |  |  |  |  |
|  |  |  |  |  |  |  |  |  |  |  |  |  |  |
|  |  |  |  |  |  |  |  |  |  |  |  |  |  |
|  |  |  |  |  |  |  |  |  |  |  |  |  |  |
|  |  |  |  |  |  |  |  |  |  |  |  |  |  |
|  |  |  |  |  |  |  |  |  |  |  |  |  |  |
|  |  |  |  |  |  |  |  |  |  |  |  |  |  |
|  |  |  |  |  |  |  |  |  |  |  |  |  |  |
|  |  |  |  |  |  |  |  |  |  |  |  |  |  |
|  |  |  |  |  |  |  |  |  |  |  |  |  |  |

## REMARKS & HAPPENINGS

# SHIP'S LOG

DATE_____ CREW_____ GUESTS_____ PAGE_____

DEPARTURE TIME_____ DESTINATION_____ ARRIVAL TIME_____

| TIME | ENGINE HOURS | PILOT INT'L | POSITION | HEADING | RPM | SPEED | ENGINE TEMP | AMPS | OIL | FUEL TANK | BARO | TEMP | WIND |
|------|------|------|------|------|------|------|------|------|------|------|------|------|------|
|  |  |  |  |  |  |  |  |  |  |  |  |  |  |
|  |  |  |  |  |  |  |  |  |  |  |  |  |  |
|  |  |  |  |  |  |  |  |  |  |  |  |  |  |
|  |  |  |  |  |  |  |  |  |  |  |  |  |  |
|  |  |  |  |  |  |  |  |  |  |  |  |  |  |
|  |  |  |  |  |  |  |  |  |  |  |  |  |  |
|  |  |  |  |  |  |  |  |  |  |  |  |  |  |
|  |  |  |  |  |  |  |  |  |  |  |  |  |  |
|  |  |  |  |  |  |  |  |  |  |  |  |  |  |
|  |  |  |  |  |  |  |  |  |  |  |  |  |  |
|  |  |  |  |  |  |  |  |  |  |  |  |  |  |
|  |  |  |  |  |  |  |  |  |  |  |  |  |  |
|  |  |  |  |  |  |  |  |  |  |  |  |  |  |
|  |  |  |  |  |  |  |  |  |  |  |  |  |  |
|  |  |  |  |  |  |  |  |  |  |  |  |  |  |
|  |  |  |  |  |  |  |  |  |  |  |  |  |  |

## REMARKS & HAPPENINGS

# SHIP'S LOG

DATE_____CREW_____GUESTS_____PAGE_____

DEPARTURE TIME_____DESTINATION_____ARRIVAL TIME_____

| TIME | ENGINE HOURS | PILOT INT'L | POSITION | HEADING | RPM | SPEED | ENGINE TEMP | AMPS | OIL | FUEL TANK | BARO | TEMP | WIND |
|------|------|------|------|------|------|------|------|------|------|------|------|------|------|
|  |  |  |  |  |  |  |  |  |  |  |  |  |  |
|  |  |  |  |  |  |  |  |  |  |  |  |  |  |
|  |  |  |  |  |  |  |  |  |  |  |  |  |  |
|  |  |  |  |  |  |  |  |  |  |  |  |  |  |
|  |  |  |  |  |  |  |  |  |  |  |  |  |  |
|  |  |  |  |  |  |  |  |  |  |  |  |  |  |
|  |  |  |  |  |  |  |  |  |  |  |  |  |  |
|  |  |  |  |  |  |  |  |  |  |  |  |  |  |
|  |  |  |  |  |  |  |  |  |  |  |  |  |  |
|  |  |  |  |  |  |  |  |  |  |  |  |  |  |
|  |  |  |  |  |  |  |  |  |  |  |  |  |  |
|  |  |  |  |  |  |  |  |  |  |  |  |  |  |
|  |  |  |  |  |  |  |  |  |  |  |  |  |  |
|  |  |  |  |  |  |  |  |  |  |  |  |  |  |
|  |  |  |  |  |  |  |  |  |  |  |  |  |  |

## REMARKS & HAPPENINGS

# SHIP'S LOG

DATE_____CREW_____GUESTS_____PAGE_____

DEPARTURE TIME_____DESTINATION_____ARRIVAL TIME_____

| TIME | ENGINE HOURS | PILOT INT'L | POSITION | HEADING | RPM | SPEED | ENGINE TEMP | AMPS | OIL | FUEL TANK | BARO | TEMP | WIND |
|------|------|------|------|------|------|------|------|------|------|------|------|------|------|
|  |  |  |  |  |  |  |  |  |  |  |  |  |  |
|  |  |  |  |  |  |  |  |  |  |  |  |  |  |
|  |  |  |  |  |  |  |  |  |  |  |  |  |  |
|  |  |  |  |  |  |  |  |  |  |  |  |  |  |
|  |  |  |  |  |  |  |  |  |  |  |  |  |  |
|  |  |  |  |  |  |  |  |  |  |  |  |  |  |
|  |  |  |  |  |  |  |  |  |  |  |  |  |  |
|  |  |  |  |  |  |  |  |  |  |  |  |  |  |
|  |  |  |  |  |  |  |  |  |  |  |  |  |  |
|  |  |  |  |  |  |  |  |  |  |  |  |  |  |
|  |  |  |  |  |  |  |  |  |  |  |  |  |  |
|  |  |  |  |  |  |  |  |  |  |  |  |  |  |

## REMARKS & HAPPENINGS

# SHIP'S LOG

DATE_____CREW_____GUESTS_____PAGE_____

DEPARTURE TIME_____DESTINATION_____ARRIVAL TIME_____

| TIME | ENGINE HOURS | PILOT INT'L | POSITION | HEADING | RPM | SPEED | ENGINE TEMP | AMPS | OIL | FUEL TANK | BARO | TEMP | WIND |
|------|------|------|------|------|------|------|------|------|------|------|------|------|------|
| | | | | | | | | | | | | | |
| | | | | | | | | | | | | | |
| | | | | | | | | | | | | | |
| | | | | | | | | | | | | | |
| | | | | | | | | | | | | | |
| | | | | | | | | | | | | | |
| | | | | | | | | | | | | | |
| | | | | | | | | | | | | | |
| | | | | | | | | | | | | | |
| | | | | | | | | | | | | | |
| | | | | | | | | | | | | | |
| | | | | | | | | | | | | | |
| | | | | | | | | | | | | | |
| | | | | | | | | | | | | | |
| | | | | | | | | | | | | | |
| | | | | | | | | | | | | | |
| | | | | | | | | | | | | | |

## REMARKS & HAPPENINGS

# SHIP'S LOG

DATE_____CREW_____GUESTS_____PAGE_____

DEPARTURE TIME_____DESTINATION_____ARRIVAL TIME_____

| TIME | ENGINE HOURS | PILOT INT'L | POSITION | HEADING | RPM | SPEED | ENGINE TEMP | AMPS | OIL | FUEL TANK | BARO | TEMP | WIND |
|------|------|------|------|------|------|------|------|------|------|------|------|------|------|
|  |  |  |  |  |  |  |  |  |  |  |  |  |  |
|  |  |  |  |  |  |  |  |  |  |  |  |  |  |
|  |  |  |  |  |  |  |  |  |  |  |  |  |  |
|  |  |  |  |  |  |  |  |  |  |  |  |  |  |
|  |  |  |  |  |  |  |  |  |  |  |  |  |  |
|  |  |  |  |  |  |  |  |  |  |  |  |  |  |
|  |  |  |  |  |  |  |  |  |  |  |  |  |  |
|  |  |  |  |  |  |  |  |  |  |  |  |  |  |
|  |  |  |  |  |  |  |  |  |  |  |  |  |  |
|  |  |  |  |  |  |  |  |  |  |  |  |  |  |
|  |  |  |  |  |  |  |  |  |  |  |  |  |  |
|  |  |  |  |  |  |  |  |  |  |  |  |  |  |

## REMARKS & HAPPENINGS

# SHIP'S LOG

DATE_____CREW_____GUESTS_____PAGE_____

DEPARTURE TIME_____DESTINATION_____ARRIVAL TIME_____

| TIME | ENGINE HOURS | PILOT INT'L | POSITION | HEADING | RPM | SPEED | ENGINE TEMP | AMPS | OIL | FUEL TANK | BARO | TEMP | WIND |
|------|------|------|------|------|------|------|------|------|------|------|------|------|------|
| | | | | | | | | | | | | | |
| | | | | | | | | | | | | | |
| | | | | | | | | | | | | | |
| | | | | | | | | | | | | | |
| | | | | | | | | | | | | | |
| | | | | | | | | | | | | | |
| | | | | | | | | | | | | | |
| | | | | | | | | | | | | | |
| | | | | | | | | | | | | | |
| | | | | | | | | | | | | | |
| | | | | | | | | | | | | | |
| | | | | | | | | | | | | | |
| | | | | | | | | | | | | | |
| | | | | | | | | | | | | | |
| | | | | | | | | | | | | | |
| | | | | | | | | | | | | | |

## REMARKS & HAPPENINGS

# SHIP'S LOG

DATE_____CREW_____GUESTS_____PAGE_____

DEPARTURE TIME_____DESTINATION_____ARRIVAL TIME_____

| TIME | ENGINE HOURS | PILOT INT'L | POSITION | HEADING | RPM | SPEED | ENGINE TEMP | AMPS | OIL | FUEL TANK | BARO | TEMP | WIND |
|------|------|------|------|------|------|------|------|------|------|------|------|------|------|
|  |  |  |  |  |  |  |  |  |  |  |  |  |  |
|  |  |  |  |  |  |  |  |  |  |  |  |  |  |
|  |  |  |  |  |  |  |  |  |  |  |  |  |  |
|  |  |  |  |  |  |  |  |  |  |  |  |  |  |
|  |  |  |  |  |  |  |  |  |  |  |  |  |  |
|  |  |  |  |  |  |  |  |  |  |  |  |  |  |
|  |  |  |  |  |  |  |  |  |  |  |  |  |  |
|  |  |  |  |  |  |  |  |  |  |  |  |  |  |
|  |  |  |  |  |  |  |  |  |  |  |  |  |  |
|  |  |  |  |  |  |  |  |  |  |  |  |  |  |
|  |  |  |  |  |  |  |  |  |  |  |  |  |  |
|  |  |  |  |  |  |  |  |  |  |  |  |  |  |
|  |  |  |  |  |  |  |  |  |  |  |  |  |  |

## REMARKS & HAPPENINGS

# SHIP'S LOG

DATE_____CREW_____GUESTS_____PAGE_____

DEPARTURE TIME_____DESTINATION_____ARRIVAL TIME_____

| TIME | ENGINE HOURS | PILOT INT'L | POSITION | HEADING | RPM | SPEED | ENGINE TEMP | AMPS | OIL | FUEL TANK | BARO | TEMP | WIND |
|------|------|------|------|------|------|------|------|------|------|------|------|------|------|
|  |  |  |  |  |  |  |  |  |  |  |  |  |  |
|  |  |  |  |  |  |  |  |  |  |  |  |  |  |
|  |  |  |  |  |  |  |  |  |  |  |  |  |  |
|  |  |  |  |  |  |  |  |  |  |  |  |  |  |
|  |  |  |  |  |  |  |  |  |  |  |  |  |  |
|  |  |  |  |  |  |  |  |  |  |  |  |  |  |
|  |  |  |  |  |  |  |  |  |  |  |  |  |  |
|  |  |  |  |  |  |  |  |  |  |  |  |  |  |
|  |  |  |  |  |  |  |  |  |  |  |  |  |  |
|  |  |  |  |  |  |  |  |  |  |  |  |  |  |
|  |  |  |  |  |  |  |  |  |  |  |  |  |  |
|  |  |  |  |  |  |  |  |  |  |  |  |  |  |
|  |  |  |  |  |  |  |  |  |  |  |  |  |  |

## REMARKS & HAPPENINGS

# SHIP'S LOG

DATE_____ CREW_____ GUESTS_____ PAGE_____

DEPARTURE TIME_____ DESTINATION_____ ARRIVAL TIME_____

| TIME | ENGINE HOURS | PILOT INT'L | POSITION | HEADING | RPM | SPEED | ENGINE TEMP | AMPS | OIL | FUEL TANK | BARO | TEMP | WIND |
|------|------|------|------|------|------|------|------|------|------|------|------|------|------|
| | | | | | | | | | | | | | |
| | | | | | | | | | | | | | |
| | | | | | | | | | | | | | |
| | | | | | | | | | | | | | |
| | | | | | | | | | | | | | |
| | | | | | | | | | | | | | |
| | | | | | | | | | | | | | |
| | | | | | | | | | | | | | |
| | | | | | | | | | | | | | |
| | | | | | | | | | | | | | |
| | | | | | | | | | | | | | |
| | | | | | | | | | | | | | |
| | | | | | | | | | | | | | |

## REMARKS & HAPPENINGS

# SHIP'S LOG

DATE_____CREW_____GUESTS_____PAGE_____

DEPARTURE TIME_____DESTINATION_____ARRIVAL TIME_____

| TIME | ENGINE HOURS | PILOT INT'L | POSITION | HEADING | RPM | SPEED | ENGINE TEMP | AMPS | OIL | FUEL TANK | BARO | TEMP | WIND |
|------|------|------|------|------|------|------|------|------|------|------|------|------|------|
| | | | | | | | | | | | | | |
| | | | | | | | | | | | | | |
| | | | | | | | | | | | | | |
| | | | | | | | | | | | | | |
| | | | | | | | | | | | | | |
| | | | | | | | | | | | | | |
| | | | | | | | | | | | | | |
| | | | | | | | | | | | | | |
| | | | | | | | | | | | | | |
| | | | | | | | | | | | | | |
| | | | | | | | | | | | | | |
| | | | | | | | | | | | | | |
| | | | | | | | | | | | | | |
| | | | | | | | | | | | | | |
| | | | | | | | | | | | | | |
| | | | | | | | | | | | | | |
| | | | | | | | | | | | | | |

## REMARKS & HAPPENINGS

# SHIP'S LOG

DATE_____CREW_____GUESTS_____PAGE_____

DEPARTURE TIME_____DESTINATION_____ARRIVAL TIME_____

| TIME | ENGINE HOURS | PILOT INT'L | POSITION | HEADING | RPM | SPEED | ENGINE TEMP | AMPS | OIL | FUEL TANK | BARO | TEMP | WIND |
|------|------|------|------|------|------|------|------|------|------|------|------|------|------|
|  |  |  |  |  |  |  |  |  |  |  |  |  |  |
|  |  |  |  |  |  |  |  |  |  |  |  |  |  |
|  |  |  |  |  |  |  |  |  |  |  |  |  |  |
|  |  |  |  |  |  |  |  |  |  |  |  |  |  |
|  |  |  |  |  |  |  |  |  |  |  |  |  |  |
|  |  |  |  |  |  |  |  |  |  |  |  |  |  |
|  |  |  |  |  |  |  |  |  |  |  |  |  |  |
|  |  |  |  |  |  |  |  |  |  |  |  |  |  |
|  |  |  |  |  |  |  |  |  |  |  |  |  |  |
|  |  |  |  |  |  |  |  |  |  |  |  |  |  |
|  |  |  |  |  |  |  |  |  |  |  |  |  |  |
|  |  |  |  |  |  |  |  |  |  |  |  |  |  |
|  |  |  |  |  |  |  |  |  |  |  |  |  |  |

## REMARKS & HAPPENINGS

# SHIP'S LOG

DATE_____CREW_____GUESTS_____PAGE_____

DEPARTURE TIME_____DESTINATION_____ARRIVAL TIME_____

| TIME | ENGINE HOURS | PILOT INT'L | POSITION | HEADING | RPM | SPEED | ENGINE TEMP | AMPS | OIL | FUEL TANK | BARO | TEMP | WIND |
|------|------|------|------|------|------|------|------|------|------|------|------|------|------|
|  |  |  |  |  |  |  |  |  |  |  |  |  |  |
|  |  |  |  |  |  |  |  |  |  |  |  |  |  |
|  |  |  |  |  |  |  |  |  |  |  |  |  |  |
|  |  |  |  |  |  |  |  |  |  |  |  |  |  |
|  |  |  |  |  |  |  |  |  |  |  |  |  |  |
|  |  |  |  |  |  |  |  |  |  |  |  |  |  |
|  |  |  |  |  |  |  |  |  |  |  |  |  |  |
|  |  |  |  |  |  |  |  |  |  |  |  |  |  |
|  |  |  |  |  |  |  |  |  |  |  |  |  |  |
|  |  |  |  |  |  |  |  |  |  |  |  |  |  |
|  |  |  |  |  |  |  |  |  |  |  |  |  |  |
|  |  |  |  |  |  |  |  |  |  |  |  |  |  |
|  |  |  |  |  |  |  |  |  |  |  |  |  |  |
|  |  |  |  |  |  |  |  |  |  |  |  |  |  |
|  |  |  |  |  |  |  |  |  |  |  |  |  |  |
|  |  |  |  |  |  |  |  |  |  |  |  |  |  |
|  |  |  |  |  |  |  |  |  |  |  |  |  |  |
|  |  |  |  |  |  |  |  |  |  |  |  |  |  |
|  |  |  |  |  |  |  |  |  |  |  |  |  |  |

## REMARKS & HAPPENINGS

# SHIP'S LOG

DATE_____CREW_____GUESTS_____PAGE_____

DEPARTURE TIME_____DESTINATION_____ARRIVAL TIME_____

| TIME | ENGINE HOURS | PILOT INT'L | POSITION | HEADING | RPM | SPEED | ENGINE TEMP | AMPS | OIL | FUEL TANK | BARO | TEMP | WIND |
|------|------|------|------|------|------|------|------|------|------|------|------|------|------|
|  |  |  |  |  |  |  |  |  |  |  |  |  |  |
|  |  |  |  |  |  |  |  |  |  |  |  |  |  |
|  |  |  |  |  |  |  |  |  |  |  |  |  |  |
|  |  |  |  |  |  |  |  |  |  |  |  |  |  |
|  |  |  |  |  |  |  |  |  |  |  |  |  |  |
|  |  |  |  |  |  |  |  |  |  |  |  |  |  |
|  |  |  |  |  |  |  |  |  |  |  |  |  |  |
|  |  |  |  |  |  |  |  |  |  |  |  |  |  |
|  |  |  |  |  |  |  |  |  |  |  |  |  |  |
|  |  |  |  |  |  |  |  |  |  |  |  |  |  |
|  |  |  |  |  |  |  |  |  |  |  |  |  |  |
|  |  |  |  |  |  |  |  |  |  |  |  |  |  |
|  |  |  |  |  |  |  |  |  |  |  |  |  |  |
|  |  |  |  |  |  |  |  |  |  |  |  |  |  |

## REMARKS & HAPPENINGS

# SHIP'S LOG

DATE_____CREW_____GUESTS_____PAGE_____

DEPARTURE TIME_____DESTINATION_____ARRIVAL TIME_____

| TIME | ENGINE HOURS | PILOT INT'L | POSITION | HEADING | RPM | SPEED | ENGINE TEMP | AMPS | OIL | FUEL TANK | BARO | TEMP | WIND |
|------|------|------|------|------|------|------|------|------|------|------|------|------|------|
| | | | | | | | | | | | | | |
| | | | | | | | | | | | | | |
| | | | | | | | | | | | | | |
| | | | | | | | | | | | | | |
| | | | | | | | | | | | | | |
| | | | | | | | | | | | | | |
| | | | | | | | | | | | | | |
| | | | | | | | | | | | | | |
| | | | | | | | | | | | | | |
| | | | | | | | | | | | | | |
| | | | | | | | | | | | | | |
| | | | | | | | | | | | | | |
| | | | | | | | | | | | | | |
| | | | | | | | | | | | | | |
| | | | | | | | | | | | | | |
| | | | | | | | | | | | | | |

## REMARKS & HAPPENINGS

# SHIP'S LOG

DATE_____CREW_____GUESTS_____PAGE_____

DEPARTURE TIME_____DESTINATION_____ARRIVAL TIME_____

| TIME | ENGINE HOURS | PILOT INT'L | POSITION | HEADING | RPM | SPEED | ENGINE TEMP | AMPS | OIL | FUEL TANK | BARO | TEMP | WIND |
|------|------|------|------|------|------|------|------|------|------|------|------|------|------|
|  |  |  |  |  |  |  |  |  |  |  |  |  |  |
|  |  |  |  |  |  |  |  |  |  |  |  |  |  |
|  |  |  |  |  |  |  |  |  |  |  |  |  |  |
|  |  |  |  |  |  |  |  |  |  |  |  |  |  |
|  |  |  |  |  |  |  |  |  |  |  |  |  |  |
|  |  |  |  |  |  |  |  |  |  |  |  |  |  |
|  |  |  |  |  |  |  |  |  |  |  |  |  |  |
|  |  |  |  |  |  |  |  |  |  |  |  |  |  |
|  |  |  |  |  |  |  |  |  |  |  |  |  |  |
|  |  |  |  |  |  |  |  |  |  |  |  |  |  |
|  |  |  |  |  |  |  |  |  |  |  |  |  |  |
|  |  |  |  |  |  |  |  |  |  |  |  |  |  |
|  |  |  |  |  |  |  |  |  |  |  |  |  |  |

## REMARKS & HAPPENINGS

# SHIP'S LOG

DATE_____CREW_____GUESTS_____PAGE_____

DEPARTURE TIME_____DESTINATION_____ARRIVAL TIME_____

| TIME | ENGINE HOURS | PILOT INT'L | POSITION | HEADING | RPM | SPEED | ENGINE TEMP | AMPS | OIL | FUEL TANK | BARO | TEMP | WIND |
|------|------|------|------|------|------|------|------|------|------|------|------|------|------|
|      |      |      |      |      |      |      |      |      |      |      |      |      |      |
|      |      |      |      |      |      |      |      |      |      |      |      |      |      |
|      |      |      |      |      |      |      |      |      |      |      |      |      |      |
|      |      |      |      |      |      |      |      |      |      |      |      |      |      |
|      |      |      |      |      |      |      |      |      |      |      |      |      |      |
|      |      |      |      |      |      |      |      |      |      |      |      |      |      |
|      |      |      |      |      |      |      |      |      |      |      |      |      |      |
|      |      |      |      |      |      |      |      |      |      |      |      |      |      |
|      |      |      |      |      |      |      |      |      |      |      |      |      |      |
|      |      |      |      |      |      |      |      |      |      |      |      |      |      |
|      |      |      |      |      |      |      |      |      |      |      |      |      |      |
|      |      |      |      |      |      |      |      |      |      |      |      |      |      |
|      |      |      |      |      |      |      |      |      |      |      |      |      |      |
|      |      |      |      |      |      |      |      |      |      |      |      |      |      |
|      |      |      |      |      |      |      |      |      |      |      |      |      |      |

## REMARKS & HAPPENINGS

# SHIP'S LOG

DATE_____CREW_____GUESTS_____PAGE_____

DEPARTURE TIME_____DESTINATION_____ARRIVAL TIME_____

| TIME | ENGINE HOURS | PILOT INT'L | POSITION | HEADING | RPM | SPEED | ENGINE TEMP | AMPS | OIL | FUEL TANK | BARO | TEMP | WIND |
|------|------|------|------|------|------|------|------|------|------|------|------|------|------|
|  |  |  |  |  |  |  |  |  |  |  |  |  |  |
|  |  |  |  |  |  |  |  |  |  |  |  |  |  |
|  |  |  |  |  |  |  |  |  |  |  |  |  |  |
|  |  |  |  |  |  |  |  |  |  |  |  |  |  |
|  |  |  |  |  |  |  |  |  |  |  |  |  |  |
|  |  |  |  |  |  |  |  |  |  |  |  |  |  |
|  |  |  |  |  |  |  |  |  |  |  |  |  |  |
|  |  |  |  |  |  |  |  |  |  |  |  |  |  |
|  |  |  |  |  |  |  |  |  |  |  |  |  |  |
|  |  |  |  |  |  |  |  |  |  |  |  |  |  |
|  |  |  |  |  |  |  |  |  |  |  |  |  |  |
|  |  |  |  |  |  |  |  |  |  |  |  |  |  |
|  |  |  |  |  |  |  |  |  |  |  |  |  |  |
|  |  |  |  |  |  |  |  |  |  |  |  |  |  |

## REMARKS & HAPPENINGS

# SHIP'S LOG

DATE_____CREW_____GUESTS_____PAGE_____

DEPARTURE TIME_____DESTINATION_____ARRIVAL TIME_____

| TIME | ENGINE HOURS | PILOT INT'L | POSITION | HEADING | RPM | SPEED | ENGINE TEMP | AMPS | OIL | FUEL TANK | BARO | TEMP | WIND |
|------|------|------|------|------|------|------|------|------|------|------|------|------|------|
| | | | | | | | | | | | | | |
| | | | | | | | | | | | | | |
| | | | | | | | | | | | | | |
| | | | | | | | | | | | | | |
| | | | | | | | | | | | | | |
| | | | | | | | | | | | | | |
| | | | | | | | | | | | | | |
| | | | | | | | | | | | | | |
| | | | | | | | | | | | | | |
| | | | | | | | | | | | | | |
| | | | | | | | | | | | | | |
| | | | | | | | | | | | | | |
| | | | | | | | | | | | | | |
| | | | | | | | | | | | | | |
| | | | | | | | | | | | | | |
| | | | | | | | | | | | | | |
| | | | | | | | | | | | | | |

## REMARKS & HAPPENINGS

# SHIP'S LOG

DATE_____CREW_____GUESTS_____PAGE_____

DEPARTURE TIME_____DESTINATION_____ARRIVAL TIME_____

| TIME | ENGINE HOURS | PILOT INT'L | POSITION | HEADING | RPM | SPEED | ENGINE TEMP | AMPS | OIL | FUEL TANK | BARO | TEMP | WIND |
|------|------|------|------|------|------|------|------|------|------|------|------|------|------|
| | | | | | | | | | | | | | |
| | | | | | | | | | | | | | |
| | | | | | | | | | | | | | |
| | | | | | | | | | | | | | |
| | | | | | | | | | | | | | |
| | | | | | | | | | | | | | |
| | | | | | | | | | | | | | |
| | | | | | | | | | | | | | |
| | | | | | | | | | | | | | |
| | | | | | | | | | | | | | |
| | | | | | | | | | | | | | |
| | | | | | | | | | | | | | |
| | | | | | | | | | | | | | |
| | | | | | | | | | | | | | |
| | | | | | | | | | | | | | |

## REMARKS & HAPPENINGS

# SHIP'S LOG

DATE_____CREW_____GUESTS_____PAGE_____

DEPARTURE TIME_____DESTINATION_____ARRIVAL TIME_____

| TIME | ENGINE HOURS | PILOT INT'L | POSITION | HEADING | RPM | SPEED | ENGINE TEMP | AMPS | OIL | FUEL TANK | BARO | TEMP | WIND |
|------|------|------|------|------|------|------|------|------|------|------|------|------|------|
| | | | | | | | | | | | | | |
| | | | | | | | | | | | | | |
| | | | | | | | | | | | | | |
| | | | | | | | | | | | | | |
| | | | | | | | | | | | | | |
| | | | | | | | | | | | | | |
| | | | | | | | | | | | | | |
| | | | | | | | | | | | | | |
| | | | | | | | | | | | | | |
| | | | | | | | | | | | | | |
| | | | | | | | | | | | | | |
| | | | | | | | | | | | | | |
| | | | | | | | | | | | | | |
| | | | | | | | | | | | | | |
| | | | | | | | | | | | | | |
| | | | | | | | | | | | | | |
| | | | | | | | | | | | | | |

## REMARKS & HAPPENINGS

# SHIP'S LOG

DATE_____CREW_____GUESTS_____PAGE_____

DEPARTURE TIME_____DESTINATION_____ARRIVAL TIME_____

| TIME | ENGINE HOURS | PILOT INT'L | POSITION | HEADING | RPM | SPEED | ENGINE TEMP | AMPS | OIL | FUEL TANK | BARO | TEMP | WIND |
|------|------|------|------|------|------|------|------|------|------|------|------|------|------|
|  |  |  |  |  |  |  |  |  |  |  |  |  |  |
|  |  |  |  |  |  |  |  |  |  |  |  |  |  |
|  |  |  |  |  |  |  |  |  |  |  |  |  |  |
|  |  |  |  |  |  |  |  |  |  |  |  |  |  |
|  |  |  |  |  |  |  |  |  |  |  |  |  |  |
|  |  |  |  |  |  |  |  |  |  |  |  |  |  |
|  |  |  |  |  |  |  |  |  |  |  |  |  |  |
|  |  |  |  |  |  |  |  |  |  |  |  |  |  |
|  |  |  |  |  |  |  |  |  |  |  |  |  |  |
|  |  |  |  |  |  |  |  |  |  |  |  |  |  |
|  |  |  |  |  |  |  |  |  |  |  |  |  |  |
|  |  |  |  |  |  |  |  |  |  |  |  |  |  |
|  |  |  |  |  |  |  |  |  |  |  |  |  |  |

## REMARKS & HAPPENINGS

# SHIP'S LOG

DATE_____CREW_____GUESTS_____PAGE_____

DEPARTURE TIME_____DESTINATION_____ARRIVAL TIME_____

| TIME | ENGINE HOURS | PILOT INT'L | POSITION | HEADING | RPM | SPEED | ENGINE TEMP | AMPS | OIL | FUEL TANK | BARO | TEMP | WIND |
|------|------|------|------|------|------|------|------|------|------|------|------|------|------|
|  |  |  |  |  |  |  |  |  |  |  |  |  |  |
|  |  |  |  |  |  |  |  |  |  |  |  |  |  |
|  |  |  |  |  |  |  |  |  |  |  |  |  |  |
|  |  |  |  |  |  |  |  |  |  |  |  |  |  |
|  |  |  |  |  |  |  |  |  |  |  |  |  |  |
|  |  |  |  |  |  |  |  |  |  |  |  |  |  |
|  |  |  |  |  |  |  |  |  |  |  |  |  |  |
|  |  |  |  |  |  |  |  |  |  |  |  |  |  |
|  |  |  |  |  |  |  |  |  |  |  |  |  |  |
|  |  |  |  |  |  |  |  |  |  |  |  |  |  |
|  |  |  |  |  |  |  |  |  |  |  |  |  |  |
|  |  |  |  |  |  |  |  |  |  |  |  |  |  |
|  |  |  |  |  |  |  |  |  |  |  |  |  |  |
|  |  |  |  |  |  |  |  |  |  |  |  |  |  |

## REMARKS & HAPPENINGS

# SHIP'S LOG

DATE_____CREW_____GUESTS_____PAGE_____

DEPARTURE TIME_____DESTINATION_____ARRIVAL TIME_____

| TIME | ENGINE HOURS | PILOT INT'L | POSITION | HEADING | RPM | SPEED | ENGINE TEMP | AMPS | OIL | FUEL TANK | BARO | TEMP | WIND |
|------|------|------|------|------|------|------|------|------|------|------|------|------|------|
|  |  |  |  |  |  |  |  |  |  |  |  |  |  |
|  |  |  |  |  |  |  |  |  |  |  |  |  |  |
|  |  |  |  |  |  |  |  |  |  |  |  |  |  |
|  |  |  |  |  |  |  |  |  |  |  |  |  |  |
|  |  |  |  |  |  |  |  |  |  |  |  |  |  |
|  |  |  |  |  |  |  |  |  |  |  |  |  |  |
|  |  |  |  |  |  |  |  |  |  |  |  |  |  |
|  |  |  |  |  |  |  |  |  |  |  |  |  |  |
|  |  |  |  |  |  |  |  |  |  |  |  |  |  |
|  |  |  |  |  |  |  |  |  |  |  |  |  |  |
|  |  |  |  |  |  |  |  |  |  |  |  |  |  |
|  |  |  |  |  |  |  |  |  |  |  |  |  |  |

## REMARKS & HAPPENINGS

# SHIP'S LOG

DATE_____CREW_____GUESTS_____PAGE_____

DEPARTURE TIME_____DESTINATION_____ARRIVAL TIME_____

| TIME | ENGINE HOURS | PILOT INT'L | POSITION | HEADING | RPM | SPEED | ENGINE TEMP | AMPS | OIL | FUEL TANK | BARO | TEMP | WIND |
|------|------|------|------|------|------|------|------|------|------|------|------|------|------|
| | | | | | | | | | | | | | |
| | | | | | | | | | | | | | |
| | | | | | | | | | | | | | |
| | | | | | | | | | | | | | |
| | | | | | | | | | | | | | |
| | | | | | | | | | | | | | |
| | | | | | | | | | | | | | |
| | | | | | | | | | | | | | |
| | | | | | | | | | | | | | |
| | | | | | | | | | | | | | |
| | | | | | | | | | | | | | |
| | | | | | | | | | | | | | |
| | | | | | | | | | | | | | |
| | | | | | | | | | | | | | |
| | | | | | | | | | | | | | |

## REMARKS & HAPPENINGS

# SHIP'S LOG

DATE_____CREW_____GUESTS_____PAGE_____

DEPARTURE TIME_____DESTINATION_____ARRIVAL TIME_____

| TIME | ENGINE HOURS | PILOT INT'L | POSITION | HEADING | RPM | SPEED | ENGINE TEMP | AMPS | OIL | FUEL TANK | BARO | TEMP | WIND |
|------|------|------|------|------|------|------|------|------|------|------|------|------|------|
| | | | | | | | | | | | | | |
| | | | | | | | | | | | | | |
| | | | | | | | | | | | | | |
| | | | | | | | | | | | | | |
| | | | | | | | | | | | | | |
| | | | | | | | | | | | | | |
| | | | | | | | | | | | | | |
| | | | | | | | | | | | | | |
| | | | | | | | | | | | | | |
| | | | | | | | | | | | | | |
| | | | | | | | | | | | | | |

## REMARKS & HAPPENINGS

# SHIP'S LOG

DATE_____CREW_____GUESTS_____PAGE_____

DEPARTURE TIME_____DESTINATION_____ARRIVAL TIME_____

| TIME | ENGINE HOURS | PILOT INT'L | POSITION | HEADING | RPM | SPEED | ENGINE TEMP | AMPS | OIL | FUEL TANK | BARO | TEMP | WIND |
|------|--------------|-------------|----------|---------|-----|-------|-------------|------|-----|-----------|------|------|------|
|      |              |             |          |         |     |       |             |      |     |           |      |      |      |
|      |              |             |          |         |     |       |             |      |     |           |      |      |      |
|      |              |             |          |         |     |       |             |      |     |           |      |      |      |
|      |              |             |          |         |     |       |             |      |     |           |      |      |      |
|      |              |             |          |         |     |       |             |      |     |           |      |      |      |
|      |              |             |          |         |     |       |             |      |     |           |      |      |      |
|      |              |             |          |         |     |       |             |      |     |           |      |      |      |
|      |              |             |          |         |     |       |             |      |     |           |      |      |      |
|      |              |             |          |         |     |       |             |      |     |           |      |      |      |
|      |              |             |          |         |     |       |             |      |     |           |      |      |      |
|      |              |             |          |         |     |       |             |      |     |           |      |      |      |
|      |              |             |          |         |     |       |             |      |     |           |      |      |      |
|      |              |             |          |         |     |       |             |      |     |           |      |      |      |
|      |              |             |          |         |     |       |             |      |     |           |      |      |      |

## REMARKS & HAPPENINGS

# SHIP'S LOG

DATE_____CREW_____GUESTS_____PAGE_____

DEPARTURE TIME_____DESTINATION_____ARRIVAL TIME_____

| TIME | ENGINE HOURS | PILOT INT'L | POSITION | HEADING | RPM | SPEED | ENGINE TEMP | AMPS | OIL | FUEL TANK | BARO | TEMP | WIND |
|------|------|------|------|------|------|------|------|------|------|------|------|------|------|
|  |  |  |  |  |  |  |  |  |  |  |  |  |  |
|  |  |  |  |  |  |  |  |  |  |  |  |  |  |
|  |  |  |  |  |  |  |  |  |  |  |  |  |  |
|  |  |  |  |  |  |  |  |  |  |  |  |  |  |
|  |  |  |  |  |  |  |  |  |  |  |  |  |  |
|  |  |  |  |  |  |  |  |  |  |  |  |  |  |
|  |  |  |  |  |  |  |  |  |  |  |  |  |  |
|  |  |  |  |  |  |  |  |  |  |  |  |  |  |
|  |  |  |  |  |  |  |  |  |  |  |  |  |  |
|  |  |  |  |  |  |  |  |  |  |  |  |  |  |
|  |  |  |  |  |  |  |  |  |  |  |  |  |  |

## REMARKS & HAPPENINGS

# SHIP'S LOG

DATE_____CREW_____GUESTS_____PAGE_____

DEPARTURE TIME_____DESTINATION_____ARRIVAL TIME_____

| TIME | ENGINE HOURS | PILOT INT'L | POSITION | HEADING | RPM | SPEED | ENGINE TEMP | AMPS | OIL | FUEL TANK | BARO | TEMP | WIND |
|------|------|------|------|------|------|------|------|------|------|------|------|------|------|
| | | | | | | | | | | | | | |
| | | | | | | | | | | | | | |
| | | | | | | | | | | | | | |
| | | | | | | | | | | | | | |
| | | | | | | | | | | | | | |
| | | | | | | | | | | | | | |
| | | | | | | | | | | | | | |
| | | | | | | | | | | | | | |
| | | | | | | | | | | | | | |
| | | | | | | | | | | | | | |
| | | | | | | | | | | | | | |
| | | | | | | | | | | | | | |
| | | | | | | | | | | | | | |
| | | | | | | | | | | | | | |
| | | | | | | | | | | | | | |
| | | | | | | | | | | | | | |
| | | | | | | | | | | | | | |
| | | | | | | | | | | | | | |

## REMARKS & HAPPENINGS

# SHIP'S LOG

DATE_____CREW_____GUESTS_____PAGE_____

DEPARTURE TIME_____DESTINATION_____ARRIVAL TIME_____

| TIME | ENGINE HOURS | PILOT INT'L | POSITION | HEADING | RPM | SPEED | ENGINE TEMP | AMPS | OIL | FUEL TANK | BARO | TEMP | WIND |
|------|--------------|-------------|----------|---------|-----|-------|-------------|------|-----|-----------|------|------|------|
|      |              |             |          |         |     |       |             |      |     |           |      |      |      |
|      |              |             |          |         |     |       |             |      |     |           |      |      |      |
|      |              |             |          |         |     |       |             |      |     |           |      |      |      |
|      |              |             |          |         |     |       |             |      |     |           |      |      |      |
|      |              |             |          |         |     |       |             |      |     |           |      |      |      |
|      |              |             |          |         |     |       |             |      |     |           |      |      |      |
|      |              |             |          |         |     |       |             |      |     |           |      |      |      |
|      |              |             |          |         |     |       |             |      |     |           |      |      |      |
|      |              |             |          |         |     |       |             |      |     |           |      |      |      |
|      |              |             |          |         |     |       |             |      |     |           |      |      |      |
|      |              |             |          |         |     |       |             |      |     |           |      |      |      |
|      |              |             |          |         |     |       |             |      |     |           |      |      |      |
|      |              |             |          |         |     |       |             |      |     |           |      |      |      |

## REMARKS & HAPPENINGS

# SHIP'S LOG

DATE_____CREW_____GUESTS_____PAGE_____

DEPARTURE TIME_____DESTINATION_____ARRIVAL TIME_____

| TIME | ENGINE HOURS | PILOT INT'L | POSITION | HEADING | RPM | SPEED | ENGINE TEMP | AMPS | OIL | FUEL TANK | BARO | TEMP | WIND |
|------|------|------|------|------|------|------|------|------|------|------|------|------|------|
| | | | | | | | | | | | | | |
| | | | | | | | | | | | | | |
| | | | | | | | | | | | | | |
| | | | | | | | | | | | | | |
| | | | | | | | | | | | | | |
| | | | | | | | | | | | | | |
| | | | | | | | | | | | | | |
| | | | | | | | | | | | | | |
| | | | | | | | | | | | | | |
| | | | | | | | | | | | | | |
| | | | | | | | | | | | | | |
| | | | | | | | | | | | | | |
| | | | | | | | | | | | | | |
| | | | | | | | | | | | | | |
| | | | | | | | | | | | | | |
| | | | | | | | | | | | | | |

## REMARKS & HAPPENINGS

# SHIP'S LOG

DATE_____CREW_____GUESTS_____PAGE_____

DEPARTURE TIME_____DESTINATION_____ARRIVAL TIME_____

| TIME | ENGINE HOURS | PILOT INT'L | POSITION | HEADING | RPM | SPEED | ENGINE TEMP | AMPS | OIL | FUEL TANK | BARO | TEMP | WIND |
|------|------|------|------|------|------|------|------|------|------|------|------|------|------|
|  |  |  |  |  |  |  |  |  |  |  |  |  |  |
|  |  |  |  |  |  |  |  |  |  |  |  |  |  |
|  |  |  |  |  |  |  |  |  |  |  |  |  |  |
|  |  |  |  |  |  |  |  |  |  |  |  |  |  |
|  |  |  |  |  |  |  |  |  |  |  |  |  |  |
|  |  |  |  |  |  |  |  |  |  |  |  |  |  |
|  |  |  |  |  |  |  |  |  |  |  |  |  |  |
|  |  |  |  |  |  |  |  |  |  |  |  |  |  |
|  |  |  |  |  |  |  |  |  |  |  |  |  |  |
|  |  |  |  |  |  |  |  |  |  |  |  |  |  |
|  |  |  |  |  |  |  |  |  |  |  |  |  |  |
|  |  |  |  |  |  |  |  |  |  |  |  |  |  |
|  |  |  |  |  |  |  |  |  |  |  |  |  |  |

## REMARKS & HAPPENINGS

# SHIP'S LOG

DATE_____CREW_____GUESTS_____PAGE_____

DEPARTURE TIME_____DESTINATION_____ARRIVAL TIME_____

| TIME | ENGINE HOURS | PILOT INT'L | POSITION | HEADING | RPM | SPEED | ENGINE TEMP | AMPS | OIL | FUEL TANK | BARO | TEMP | WIND |
|------|------|------|------|------|------|------|------|------|------|------|------|------|------|
|  |  |  |  |  |  |  |  |  |  |  |  |  |  |
|  |  |  |  |  |  |  |  |  |  |  |  |  |  |
|  |  |  |  |  |  |  |  |  |  |  |  |  |  |
|  |  |  |  |  |  |  |  |  |  |  |  |  |  |
|  |  |  |  |  |  |  |  |  |  |  |  |  |  |
|  |  |  |  |  |  |  |  |  |  |  |  |  |  |
|  |  |  |  |  |  |  |  |  |  |  |  |  |  |
|  |  |  |  |  |  |  |  |  |  |  |  |  |  |
|  |  |  |  |  |  |  |  |  |  |  |  |  |  |
|  |  |  |  |  |  |  |  |  |  |  |  |  |  |
|  |  |  |  |  |  |  |  |  |  |  |  |  |  |
|  |  |  |  |  |  |  |  |  |  |  |  |  |  |
|  |  |  |  |  |  |  |  |  |  |  |  |  |  |
|  |  |  |  |  |  |  |  |  |  |  |  |  |  |
|  |  |  |  |  |  |  |  |  |  |  |  |  |  |

## REMARKS & HAPPENINGS

# SHIP'S LOG

DATE_____CREW_____GUESTS_____PAGE_____

DEPARTURE TIME_____DESTINATION_____ARRIVAL TIME_____

| TIME | ENGINE HOURS | PILOT INT'L | POSITION | HEADING | RPM | SPEED | ENGINE TEMP | AMPS | OIL | FUEL TANK | BARO | TEMP | WIND |
|------|------|------|------|------|------|------|------|------|------|------|------|------|------|
|  |  |  |  |  |  |  |  |  |  |  |  |  |  |
|  |  |  |  |  |  |  |  |  |  |  |  |  |  |
|  |  |  |  |  |  |  |  |  |  |  |  |  |  |
|  |  |  |  |  |  |  |  |  |  |  |  |  |  |
|  |  |  |  |  |  |  |  |  |  |  |  |  |  |
|  |  |  |  |  |  |  |  |  |  |  |  |  |  |
|  |  |  |  |  |  |  |  |  |  |  |  |  |  |
|  |  |  |  |  |  |  |  |  |  |  |  |  |  |
|  |  |  |  |  |  |  |  |  |  |  |  |  |  |
|  |  |  |  |  |  |  |  |  |  |  |  |  |  |
|  |  |  |  |  |  |  |  |  |  |  |  |  |  |
|  |  |  |  |  |  |  |  |  |  |  |  |  |  |
|  |  |  |  |  |  |  |  |  |  |  |  |  |  |
|  |  |  |  |  |  |  |  |  |  |  |  |  |  |
|  |  |  |  |  |  |  |  |  |  |  |  |  |  |

## REMARKS & HAPPENINGS

# SHIP'S LOG

DATE_____CREW_____GUESTS_____PAGE_____

DEPARTURE TIME_____DESTINATION_____ARRIVAL TIME_____

| TIME | ENGINE HOURS | PILOT INT'L | POSITION | HEADING | RPM | SPEED | ENGINE TEMP | AMPS | OIL | FUEL TANK | BARO | TEMP | WIND |
|------|------|------|------|------|------|------|------|------|------|------|------|------|------|
| | | | | | | | | | | | | | |
| | | | | | | | | | | | | | |
| | | | | | | | | | | | | | |
| | | | | | | | | | | | | | |
| | | | | | | | | | | | | | |
| | | | | | | | | | | | | | |
| | | | | | | | | | | | | | |
| | | | | | | | | | | | | | |
| | | | | | | | | | | | | | |
| | | | | | | | | | | | | | |
| | | | | | | | | | | | | | |
| | | | | | | | | | | | | | |
| | | | | | | | | | | | | | |
| | | | | | | | | | | | | | |
| | | | | | | | | | | | | | |

## REMARKS & HAPPENINGS

# SHIP'S LOG

DATE_____CREW_____GUESTS_____PAGE_____

DEPARTURE TIME_____DESTINATION_____ARRIVAL TIME_____

| TIME | ENGINE HOURS | PILOT INT'L | POSITION | HEADING | RPM | SPEED | ENGINE TEMP | AMPS | OIL | FUEL TANK | BARO | TEMP | WIND |
|------|------|------|------|------|------|------|------|------|------|------|------|------|------|
| | | | | | | | | | | | | | |
| | | | | | | | | | | | | | |
| | | | | | | | | | | | | | |
| | | | | | | | | | | | | | |
| | | | | | | | | | | | | | |
| | | | | | | | | | | | | | |
| | | | | | | | | | | | | | |
| | | | | | | | | | | | | | |
| | | | | | | | | | | | | | |
| | | | | | | | | | | | | | |
| | | | | | | | | | | | | | |
| | | | | | | | | | | | | | |
| | | | | | | | | | | | | | |
| | | | | | | | | | | | | | |
| | | | | | | | | | | | | | |
| | | | | | | | | | | | | | |

## REMARKS & HAPPENINGS

# SHIP'S LOG

DATE_____CREW_____GUESTS_____PAGE_____

DEPARTURE TIME_____DESTINATION_____ARRIVAL TIME_____

| TIME | ENGINE HOURS | PILOT INT'L | POSITION | HEADING | RPM | SPEED | ENGINE TEMP | AMPS | OIL | FUEL TANK | BARO | TEMP | WIND |
|------|--------------|-------------|----------|---------|-----|-------|-------------|------|-----|-----------|------|------|------|
|      |              |             |          |         |     |       |             |      |     |           |      |      |      |
|      |              |             |          |         |     |       |             |      |     |           |      |      |      |
|      |              |             |          |         |     |       |             |      |     |           |      |      |      |
|      |              |             |          |         |     |       |             |      |     |           |      |      |      |
|      |              |             |          |         |     |       |             |      |     |           |      |      |      |
|      |              |             |          |         |     |       |             |      |     |           |      |      |      |
|      |              |             |          |         |     |       |             |      |     |           |      |      |      |
|      |              |             |          |         |     |       |             |      |     |           |      |      |      |
|      |              |             |          |         |     |       |             |      |     |           |      |      |      |
|      |              |             |          |         |     |       |             |      |     |           |      |      |      |
|      |              |             |          |         |     |       |             |      |     |           |      |      |      |
|      |              |             |          |         |     |       |             |      |     |           |      |      |      |
|      |              |             |          |         |     |       |             |      |     |           |      |      |      |
|      |              |             |          |         |     |       |             |      |     |           |      |      |      |
|      |              |             |          |         |     |       |             |      |     |           |      |      |      |
|      |              |             |          |         |     |       |             |      |     |           |      |      |      |
|      |              |             |          |         |     |       |             |      |     |           |      |      |      |

## REMARKS & HAPPENINGS

# SHIP'S LOG

DATE_____CREW_____GUESTS_____PAGE_____

DEPARTURE TIME_____DESTINATION_____ARRIVAL TIME_____

| TIME | ENGINE HOURS | PILOT INT'L | POSITION | HEADING | RPM | SPEED | ENGINE TEMP | AMPS | OIL | FUEL TANK | BARO | TEMP | WIND |
|------|------|------|------|------|------|------|------|------|------|------|------|------|------|
|  |  |  |  |  |  |  |  |  |  |  |  |  |  |
|  |  |  |  |  |  |  |  |  |  |  |  |  |  |
|  |  |  |  |  |  |  |  |  |  |  |  |  |  |
|  |  |  |  |  |  |  |  |  |  |  |  |  |  |
|  |  |  |  |  |  |  |  |  |  |  |  |  |  |
|  |  |  |  |  |  |  |  |  |  |  |  |  |  |
|  |  |  |  |  |  |  |  |  |  |  |  |  |  |
|  |  |  |  |  |  |  |  |  |  |  |  |  |  |
|  |  |  |  |  |  |  |  |  |  |  |  |  |  |
|  |  |  |  |  |  |  |  |  |  |  |  |  |  |
|  |  |  |  |  |  |  |  |  |  |  |  |  |  |
|  |  |  |  |  |  |  |  |  |  |  |  |  |  |
|  |  |  |  |  |  |  |  |  |  |  |  |  |  |

## REMARKS & HAPPENINGS

# SHIP'S LOG

DATE_____CREW_____GUESTS_____PAGE_____

DEPARTURE TIME_____DESTINATION_____ARRIVAL TIME_____

| TIME | ENGINE HOURS | PILOT INT'L | POSITION | HEADING | RPM | SPEED | ENGINE TEMP | AMPS | OIL | FUEL TANK | BARO | TEMP | WIND |
|------|------|------|------|------|------|------|------|------|------|------|------|------|------|
|  |  |  |  |  |  |  |  |  |  |  |  |  |  |
|  |  |  |  |  |  |  |  |  |  |  |  |  |  |
|  |  |  |  |  |  |  |  |  |  |  |  |  |  |
|  |  |  |  |  |  |  |  |  |  |  |  |  |  |
|  |  |  |  |  |  |  |  |  |  |  |  |  |  |
|  |  |  |  |  |  |  |  |  |  |  |  |  |  |
|  |  |  |  |  |  |  |  |  |  |  |  |  |  |
|  |  |  |  |  |  |  |  |  |  |  |  |  |  |
|  |  |  |  |  |  |  |  |  |  |  |  |  |  |
|  |  |  |  |  |  |  |  |  |  |  |  |  |  |
|  |  |  |  |  |  |  |  |  |  |  |  |  |  |
|  |  |  |  |  |  |  |  |  |  |  |  |  |  |
|  |  |  |  |  |  |  |  |  |  |  |  |  |  |
|  |  |  |  |  |  |  |  |  |  |  |  |  |  |
|  |  |  |  |  |  |  |  |  |  |  |  |  |  |

## REMARKS & HAPPENINGS

# SHIP'S LOG

DATE_____CREW_____GUESTS_____PAGE_____

DEPARTURE TIME_____DESTINATION_____ARRIVAL TIME_____

| TIME | ENGINE HOURS | PILOT INT'L | POSITION | HEADING | RPM | SPEED | ENGINE TEMP | AMPS | OIL | FUEL TANK | BARO | TEMP | WIND |
|---|---|---|---|---|---|---|---|---|---|---|---|---|---|
| | | | | | | | | | | | | | |
| | | | | | | | | | | | | | |
| | | | | | | | | | | | | | |
| | | | | | | | | | | | | | |
| | | | | | | | | | | | | | |
| | | | | | | | | | | | | | |
| | | | | | | | | | | | | | |
| | | | | | | | | | | | | | |
| | | | | | | | | | | | | | |
| | | | | | | | | | | | | | |
| | | | | | | | | | | | | | |
| | | | | | | | | | | | | | |
| | | | | | | | | | | | | | |

## REMARKS & HAPPENINGS

# SHIP'S LOG

DATE_____CREW_____GUESTS_____PAGE_____

DEPARTURE TIME_____DESTINATION_____ARRIVAL TIME_____

| TIME | ENGINE HOURS | PILOT INT'L | POSITION | HEADING | RPM | SPEED | ENGINE TEMP | AMPS | OIL | FUEL TANK | BARO | TEMP | WIND |
|------|------|------|------|------|------|------|------|------|------|------|------|------|------|
|  |  |  |  |  |  |  |  |  |  |  |  |  |  |
|  |  |  |  |  |  |  |  |  |  |  |  |  |  |
|  |  |  |  |  |  |  |  |  |  |  |  |  |  |
|  |  |  |  |  |  |  |  |  |  |  |  |  |  |
|  |  |  |  |  |  |  |  |  |  |  |  |  |  |
|  |  |  |  |  |  |  |  |  |  |  |  |  |  |
|  |  |  |  |  |  |  |  |  |  |  |  |  |  |
|  |  |  |  |  |  |  |  |  |  |  |  |  |  |
|  |  |  |  |  |  |  |  |  |  |  |  |  |  |
|  |  |  |  |  |  |  |  |  |  |  |  |  |  |
|  |  |  |  |  |  |  |  |  |  |  |  |  |  |
|  |  |  |  |  |  |  |  |  |  |  |  |  |  |
|  |  |  |  |  |  |  |  |  |  |  |  |  |  |
|  |  |  |  |  |  |  |  |  |  |  |  |  |  |

## REMARKS & HAPPENINGS

# SHIP'S LOG

DATE_____CREW_____GUESTS_____PAGE_____

DEPARTURE TIME_____DESTINATION_____ARRIVAL TIME_____

| TIME | ENGINE HOURS | PILOT INT'L | POSITION | HEADING | RPM | SPEED | ENGINE TEMP | AMPS | OIL | FUEL TANK | BARO | TEMP | WIND |
|---|---|---|---|---|---|---|---|---|---|---|---|---|---|
| | | | | | | | | | | | | | |
| | | | | | | | | | | | | | |
| | | | | | | | | | | | | | |
| | | | | | | | | | | | | | |
| | | | | | | | | | | | | | |
| | | | | | | | | | | | | | |
| | | | | | | | | | | | | | |
| | | | | | | | | | | | | | |
| | | | | | | | | | | | | | |
| | | | | | | | | | | | | | |
| | | | | | | | | | | | | | |
| | | | | | | | | | | | | | |
| | | | | | | | | | | | | | |
| | | | | | | | | | | | | | |
| | | | | | | | | | | | | | |
| | | | | | | | | | | | | | |

REMARKS & HAPPENINGS

# SHIP'S LOG

DATE_____CREW_____GUESTS_____PAGE_____

DEPARTURE TIME_____DESTINATION_____ARRIVAL TIME_____

| TIME | ENGINE HOURS | PILOT INT'L | POSITION | HEADING | RPM | SPEED | ENGINE TEMP | AMPS | OIL | FUEL TANK | BARO | TEMP | WIND |
|------|--------------|-------------|----------|---------|-----|-------|-------------|------|-----|-----------|------|------|------|
|      |              |             |          |         |     |       |             |      |     |           |      |      |      |
|      |              |             |          |         |     |       |             |      |     |           |      |      |      |
|      |              |             |          |         |     |       |             |      |     |           |      |      |      |
|      |              |             |          |         |     |       |             |      |     |           |      |      |      |
|      |              |             |          |         |     |       |             |      |     |           |      |      |      |
|      |              |             |          |         |     |       |             |      |     |           |      |      |      |
|      |              |             |          |         |     |       |             |      |     |           |      |      |      |
|      |              |             |          |         |     |       |             |      |     |           |      |      |      |
|      |              |             |          |         |     |       |             |      |     |           |      |      |      |
|      |              |             |          |         |     |       |             |      |     |           |      |      |      |
|      |              |             |          |         |     |       |             |      |     |           |      |      |      |
|      |              |             |          |         |     |       |             |      |     |           |      |      |      |
|      |              |             |          |         |     |       |             |      |     |           |      |      |      |
|      |              |             |          |         |     |       |             |      |     |           |      |      |      |
|      |              |             |          |         |     |       |             |      |     |           |      |      |      |

## REMARKS & HAPPENINGS

# SHIP'S LOG

DATE_____CREW_____GUESTS_____PAGE_____

DEPARTURE TIME_____DESTINATION_____ARRIVAL TIME_____

| TIME | ENGINE HOURS | PILOT INT'L | POSITION | HEADING | RPM | SPEED | ENGINE TEMP | AMPS | OIL | FUEL TANK | BARO | TEMP | WIND |
|------|------|------|------|------|------|------|------|------|------|------|------|------|------|
|  |  |  |  |  |  |  |  |  |  |  |  |  |  |
|  |  |  |  |  |  |  |  |  |  |  |  |  |  |
|  |  |  |  |  |  |  |  |  |  |  |  |  |  |
|  |  |  |  |  |  |  |  |  |  |  |  |  |  |
|  |  |  |  |  |  |  |  |  |  |  |  |  |  |
|  |  |  |  |  |  |  |  |  |  |  |  |  |  |
|  |  |  |  |  |  |  |  |  |  |  |  |  |  |
|  |  |  |  |  |  |  |  |  |  |  |  |  |  |
|  |  |  |  |  |  |  |  |  |  |  |  |  |  |
|  |  |  |  |  |  |  |  |  |  |  |  |  |  |
|  |  |  |  |  |  |  |  |  |  |  |  |  |  |
|  |  |  |  |  |  |  |  |  |  |  |  |  |  |
|  |  |  |  |  |  |  |  |  |  |  |  |  |  |
|  |  |  |  |  |  |  |  |  |  |  |  |  |  |

## REMARKS & HAPPENINGS

# SHIP'S LOG

DATE_____CREW_____GUESTS_____PAGE_____

DEPARTURE TIME_____DESTINATION_____ARRIVAL TIME_____

| TIME | ENGINE HOURS | PILOT INT'L | POSITION | HEADING | RPM | SPEED | ENGINE TEMP | AMPS | OIL | FUEL TANK | BARO | TEMP | WIND |
|------|------|------|------|------|------|------|------|------|------|------|------|------|------|
| | | | | | | | | | | | | | |
| | | | | | | | | | | | | | |
| | | | | | | | | | | | | | |
| | | | | | | | | | | | | | |
| | | | | | | | | | | | | | |
| | | | | | | | | | | | | | |
| | | | | | | | | | | | | | |
| | | | | | | | | | | | | | |
| | | | | | | | | | | | | | |
| | | | | | | | | | | | | | |
| | | | | | | | | | | | | | |
| | | | | | | | | | | | | | |
| | | | | | | | | | | | | | |
| | | | | | | | | | | | | | |
| | | | | | | | | | | | | | |
| | | | | | | | | | | | | | |
| | | | | | | | | | | | | | |
| | | | | | | | | | | | | | |
| | | | | | | | | | | | | | |

## REMARKS & HAPPENINGS

# SHIP'S LOG

DATE_____CREW_____GUESTS_____PAGE_____

DEPARTURE TIME_____DESTINATION_____ARRIVAL TIME_____

| TIME | ENGINE HOURS | PILOT INT'L | POSITION | HEADING | RPM | SPEED | ENGINE TEMP | AMPS | OIL | FUEL TANK | BARO | TEMP | WIND |
|------|------|------|------|------|------|------|------|------|------|------|------|------|------|
| | | | | | | | | | | | | | |
| | | | | | | | | | | | | | |
| | | | | | | | | | | | | | |
| | | | | | | | | | | | | | |
| | | | | | | | | | | | | | |
| | | | | | | | | | | | | | |
| | | | | | | | | | | | | | |
| | | | | | | | | | | | | | |
| | | | | | | | | | | | | | |
| | | | | | | | | | | | | | |
| | | | | | | | | | | | | | |
| | | | | | | | | | | | | | |
| | | | | | | | | | | | | | |
| | | | | | | | | | | | | | |
| | | | | | | | | | | | | | |
| | | | | | | | | | | | | | |
| | | | | | | | | | | | | | |

## REMARKS & HAPPENINGS

# SHIP'S LOG

DATE_____CREW_____GUESTS_____PAGE_____

DEPARTURE TIME_____DESTINATION_____ARRIVAL TIME_____

| TIME | ENGINE HOURS | PILOT INT'L | POSITION | HEADING | RPM | SPEED | ENGINE TEMP | AMPS | OIL | FUEL TANK | BARO | TEMP | WIND |
|------|------|------|------|------|------|------|------|------|------|------|------|------|------|
|  |  |  |  |  |  |  |  |  |  |  |  |  |  |
|  |  |  |  |  |  |  |  |  |  |  |  |  |  |
|  |  |  |  |  |  |  |  |  |  |  |  |  |  |
|  |  |  |  |  |  |  |  |  |  |  |  |  |  |
|  |  |  |  |  |  |  |  |  |  |  |  |  |  |
|  |  |  |  |  |  |  |  |  |  |  |  |  |  |
|  |  |  |  |  |  |  |  |  |  |  |  |  |  |
|  |  |  |  |  |  |  |  |  |  |  |  |  |  |
|  |  |  |  |  |  |  |  |  |  |  |  |  |  |
|  |  |  |  |  |  |  |  |  |  |  |  |  |  |
|  |  |  |  |  |  |  |  |  |  |  |  |  |  |
|  |  |  |  |  |  |  |  |  |  |  |  |  |  |
|  |  |  |  |  |  |  |  |  |  |  |  |  |  |

## REMARKS & HAPPENINGS

# SHIP'S LOG

DATE_____CREW_____GUESTS_____PAGE_____

DEPARTURE TIME_____DESTINATION_____ARRIVAL TIME_____

| TIME | ENGINE HOURS | PILOT INT'L | POSITION | HEADING | RPM | SPEED | ENGINE TEMP | AMPS | OIL | FUEL TANK | BARO | TEMP | WIND |
|---|---|---|---|---|---|---|---|---|---|---|---|---|---|
|  |  |  |  |  |  |  |  |  |  |  |  |  |  |
|  |  |  |  |  |  |  |  |  |  |  |  |  |  |
|  |  |  |  |  |  |  |  |  |  |  |  |  |  |
|  |  |  |  |  |  |  |  |  |  |  |  |  |  |
|  |  |  |  |  |  |  |  |  |  |  |  |  |  |
|  |  |  |  |  |  |  |  |  |  |  |  |  |  |
|  |  |  |  |  |  |  |  |  |  |  |  |  |  |
|  |  |  |  |  |  |  |  |  |  |  |  |  |  |
|  |  |  |  |  |  |  |  |  |  |  |  |  |  |
|  |  |  |  |  |  |  |  |  |  |  |  |  |  |
|  |  |  |  |  |  |  |  |  |  |  |  |  |  |
|  |  |  |  |  |  |  |  |  |  |  |  |  |  |
|  |  |  |  |  |  |  |  |  |  |  |  |  |  |
|  |  |  |  |  |  |  |  |  |  |  |  |  |  |

## REMARKS & HAPPENINGS

# SHIP'S LOG

DATE_____CREW_____GUESTS_____PAGE_____

DEPARTURE TIME_____DESTINATION_____ARRIVAL TIME_____

| TIME | ENGINE HOURS | PILOT INT'L | POSITION | HEADING | RPM | SPEED | ENGINE TEMP | AMPS | OIL | FUEL TANK | BARO | TEMP | WIND |
|------|------|------|------|------|------|------|------|------|------|------|------|------|------|
| | | | | | | | | | | | | | |
| | | | | | | | | | | | | | |
| | | | | | | | | | | | | | |
| | | | | | | | | | | | | | |
| | | | | | | | | | | | | | |
| | | | | | | | | | | | | | |
| | | | | | | | | | | | | | |
| | | | | | | | | | | | | | |
| | | | | | | | | | | | | | |
| | | | | | | | | | | | | | |
| | | | | | | | | | | | | | |
| | | | | | | | | | | | | | |
| | | | | | | | | | | | | | |
| | | | | | | | | | | | | | |
| | | | | | | | | | | | | | |
| | | | | | | | | | | | | | |

## REMARKS & HAPPENINGS

# SHIP'S LOG

DATE_____CREW_____GUESTS_____PAGE_____

DEPARTURE TIME_____DESTINATION_____ARRIVAL TIME_____

| TIME | ENGINE HOURS | PILOT INT'L | POSITION | HEADING | RPM | SPEED | ENGINE TEMP | AMPS | OIL | FUEL TANK | BARO | TEMP | WIND |
|------|------|------|------|------|------|------|------|------|------|------|------|------|------|
|  |  |  |  |  |  |  |  |  |  |  |  |  |  |
|  |  |  |  |  |  |  |  |  |  |  |  |  |  |
|  |  |  |  |  |  |  |  |  |  |  |  |  |  |
|  |  |  |  |  |  |  |  |  |  |  |  |  |  |
|  |  |  |  |  |  |  |  |  |  |  |  |  |  |
|  |  |  |  |  |  |  |  |  |  |  |  |  |  |
|  |  |  |  |  |  |  |  |  |  |  |  |  |  |
|  |  |  |  |  |  |  |  |  |  |  |  |  |  |
|  |  |  |  |  |  |  |  |  |  |  |  |  |  |
|  |  |  |  |  |  |  |  |  |  |  |  |  |  |
|  |  |  |  |  |  |  |  |  |  |  |  |  |  |
|  |  |  |  |  |  |  |  |  |  |  |  |  |  |
|  |  |  |  |  |  |  |  |  |  |  |  |  |  |

## REMARKS & HAPPENINGS

# SHIP'S LOG

DATE_____CREW_____GUESTS_____PAGE_____

DEPARTURE TIME_____DESTINATION_____ARRIVAL TIME_____

| TIME | ENGINE HOURS | PILOT INT'L | POSITION | HEADING | RPM | SPEED | ENGINE TEMP | AMPS | OIL | FUEL TANK | BARO | TEMP | WIND |
|------|--------------|-------------|----------|---------|-----|-------|-------------|------|-----|-----------|------|------|------|
|      |              |             |          |         |     |       |             |      |     |           |      |      |      |
|      |              |             |          |         |     |       |             |      |     |           |      |      |      |
|      |              |             |          |         |     |       |             |      |     |           |      |      |      |
|      |              |             |          |         |     |       |             |      |     |           |      |      |      |
|      |              |             |          |         |     |       |             |      |     |           |      |      |      |
|      |              |             |          |         |     |       |             |      |     |           |      |      |      |
|      |              |             |          |         |     |       |             |      |     |           |      |      |      |
|      |              |             |          |         |     |       |             |      |     |           |      |      |      |
|      |              |             |          |         |     |       |             |      |     |           |      |      |      |
|      |              |             |          |         |     |       |             |      |     |           |      |      |      |
|      |              |             |          |         |     |       |             |      |     |           |      |      |      |
|      |              |             |          |         |     |       |             |      |     |           |      |      |      |
|      |              |             |          |         |     |       |             |      |     |           |      |      |      |
|      |              |             |          |         |     |       |             |      |     |           |      |      |      |

## REMARKS & HAPPENINGS

# SHIP'S LOG

DATE_____CREW_____GUESTS_____PAGE_____

DEPARTURE TIME_____DESTINATION_____ARRIVAL TIME_____

| TIME | ENGINE HOURS | PILOT INT'L | POSITION | HEADING | RPM | SPEED | ENGINE TEMP | AMPS | OIL | FUEL TANK | BARO | TEMP | WIND |
|------|------|------|------|------|------|------|------|------|------|------|------|------|------|
| | | | | | | | | | | | | | |
| | | | | | | | | | | | | | |
| | | | | | | | | | | | | | |
| | | | | | | | | | | | | | |
| | | | | | | | | | | | | | |
| | | | | | | | | | | | | | |
| | | | | | | | | | | | | | |
| | | | | | | | | | | | | | |
| | | | | | | | | | | | | | |
| | | | | | | | | | | | | | |
| | | | | | | | | | | | | | |
| | | | | | | | | | | | | | |
| | | | | | | | | | | | | | |
| | | | | | | | | | | | | | |
| | | | | | | | | | | | | | |

## REMARKS & HAPPENINGS

# SHIP'S LOG

DATE_____CREW_____GUESTS_____PAGE_____

DEPARTURE TIME_____DESTINATION_____ARRIVAL TIME_____

| TIME | ENGINE HOURS | PILOT INT'L | POSITION | HEADING | RPM | SPEED | ENGINE TEMP | AMPS | OIL | FUEL TANK | BARO | TEMP | WIND |
|------|--------------|-------------|----------|---------|-----|-------|-------------|------|-----|-----------|------|------|------|
|      |              |             |          |         |     |       |             |      |     |           |      |      |      |
|      |              |             |          |         |     |       |             |      |     |           |      |      |      |
|      |              |             |          |         |     |       |             |      |     |           |      |      |      |
|      |              |             |          |         |     |       |             |      |     |           |      |      |      |
|      |              |             |          |         |     |       |             |      |     |           |      |      |      |
|      |              |             |          |         |     |       |             |      |     |           |      |      |      |
|      |              |             |          |         |     |       |             |      |     |           |      |      |      |
|      |              |             |          |         |     |       |             |      |     |           |      |      |      |
|      |              |             |          |         |     |       |             |      |     |           |      |      |      |
|      |              |             |          |         |     |       |             |      |     |           |      |      |      |
|      |              |             |          |         |     |       |             |      |     |           |      |      |      |
|      |              |             |          |         |     |       |             |      |     |           |      |      |      |
|      |              |             |          |         |     |       |             |      |     |           |      |      |      |
|      |              |             |          |         |     |       |             |      |     |           |      |      |      |

## REMARKS & HAPPENINGS

# SHIP'S LOG

DATE_____CREW_____GUESTS_____PAGE_____

DEPARTURE TIME_____DESTINATION_____ARRIVAL TIME_____

| TIME | ENGINE HOURS | PILOT INT'L | POSITION | HEADING | RPM | SPEED | ENGINE TEMP | AMPS | OIL | FUEL TANK | BARO | TEMP | WIND |
|------|--------------|-------------|----------|---------|-----|-------|-------------|------|-----|-----------|------|------|------|
|      |              |             |          |         |     |       |             |      |     |           |      |      |      |
|      |              |             |          |         |     |       |             |      |     |           |      |      |      |
|      |              |             |          |         |     |       |             |      |     |           |      |      |      |
|      |              |             |          |         |     |       |             |      |     |           |      |      |      |
|      |              |             |          |         |     |       |             |      |     |           |      |      |      |
|      |              |             |          |         |     |       |             |      |     |           |      |      |      |
|      |              |             |          |         |     |       |             |      |     |           |      |      |      |
|      |              |             |          |         |     |       |             |      |     |           |      |      |      |
|      |              |             |          |         |     |       |             |      |     |           |      |      |      |
|      |              |             |          |         |     |       |             |      |     |           |      |      |      |
|      |              |             |          |         |     |       |             |      |     |           |      |      |      |
|      |              |             |          |         |     |       |             |      |     |           |      |      |      |
|      |              |             |          |         |     |       |             |      |     |           |      |      |      |

## REMARKS & HAPPENINGS

# SHIP'S LOG

DATE_____CREW_____GUESTS_____PAGE_____

DEPARTURE TIME_____DESTINATION_____ARRIVAL TIME_____

| TIME | ENGINE HOURS | PILOT INT'L | POSITION | HEADING | RPM | SPEED | ENGINE TEMP | AMPS | OIL | FUEL TANK | BARO | TEMP | WIND |
|------|------|------|------|------|------|------|------|------|------|------|------|------|------|
| | | | | | | | | | | | | | |
| | | | | | | | | | | | | | |
| | | | | | | | | | | | | | |
| | | | | | | | | | | | | | |
| | | | | | | | | | | | | | |
| | | | | | | | | | | | | | |
| | | | | | | | | | | | | | |
| | | | | | | | | | | | | | |
| | | | | | | | | | | | | | |
| | | | | | | | | | | | | | |
| | | | | | | | | | | | | | |
| | | | | | | | | | | | | | |
| | | | | | | | | | | | | | |
| | | | | | | | | | | | | | |
| | | | | | | | | | | | | | |
| | | | | | | | | | | | | | |
| | | | | | | | | | | | | | |
| | | | | | | | | | | | | | |

## REMARKS & HAPPENINGS

# SHIP'S LOG

DATE_____CREW_____GUESTS_____PAGE_____

DEPARTURE TIME_____DESTINATION_____ARRIVAL TIME_____

| TIME | ENGINE HOURS | PILOT INT'L | POSITION | HEADING | RPM | SPEED | ENGINE TEMP | AMPS | OIL | FUEL TANK | BARO | TEMP | WIND |
|------|------|------|------|------|------|------|------|------|------|------|------|------|------|
|  |  |  |  |  |  |  |  |  |  |  |  |  |  |
|  |  |  |  |  |  |  |  |  |  |  |  |  |  |
|  |  |  |  |  |  |  |  |  |  |  |  |  |  |
|  |  |  |  |  |  |  |  |  |  |  |  |  |  |
|  |  |  |  |  |  |  |  |  |  |  |  |  |  |
|  |  |  |  |  |  |  |  |  |  |  |  |  |  |
|  |  |  |  |  |  |  |  |  |  |  |  |  |  |
|  |  |  |  |  |  |  |  |  |  |  |  |  |  |
|  |  |  |  |  |  |  |  |  |  |  |  |  |  |
|  |  |  |  |  |  |  |  |  |  |  |  |  |  |
|  |  |  |  |  |  |  |  |  |  |  |  |  |  |
|  |  |  |  |  |  |  |  |  |  |  |  |  |  |
|  |  |  |  |  |  |  |  |  |  |  |  |  |  |

## REMARKS & HAPPENINGS

# SHIP'S LOG

DATE_____ CREW_____ GUESTS_____ PAGE_____

DEPARTURE TIME_____ DESTINATION_____ ARRIVAL TIME_____

| TIME | ENGINE HOURS | PILOT INT'L | POSITION | HEADING | RPM | SPEED | ENGINE TEMP | AMPS | OIL | FUEL TANK | BARO | TEMP | WIND |
|------|------|------|------|------|------|------|------|------|------|------|------|------|------|
| | | | | | | | | | | | | | |
| | | | | | | | | | | | | | |
| | | | | | | | | | | | | | |
| | | | | | | | | | | | | | |
| | | | | | | | | | | | | | |
| | | | | | | | | | | | | | |
| | | | | | | | | | | | | | |
| | | | | | | | | | | | | | |
| | | | | | | | | | | | | | |
| | | | | | | | | | | | | | |
| | | | | | | | | | | | | | |
| | | | | | | | | | | | | | |
| | | | | | | | | | | | | | |
| | | | | | | | | | | | | | |
| | | | | | | | | | | | | | |
| | | | | | | | | | | | | | |

## REMARKS & HAPPENINGS

# SHIP'S LOG

DATE_____CREW_____GUESTS_____PAGE_____

DEPARTURE TIME_____DESTINATION_____ARRIVAL TIME_____

| TIME | ENGINE HOURS | PILOT INT'L | POSITION | HEADING | RPM | SPEED | ENGINE TEMP | AMPS | OIL | FUEL TANK | BARO | TEMP | WIND |
|------|------|------|------|------|------|------|------|------|------|------|------|------|------|
|  |  |  |  |  |  |  |  |  |  |  |  |  |  |
|  |  |  |  |  |  |  |  |  |  |  |  |  |  |
|  |  |  |  |  |  |  |  |  |  |  |  |  |  |
|  |  |  |  |  |  |  |  |  |  |  |  |  |  |
|  |  |  |  |  |  |  |  |  |  |  |  |  |  |
|  |  |  |  |  |  |  |  |  |  |  |  |  |  |
|  |  |  |  |  |  |  |  |  |  |  |  |  |  |
|  |  |  |  |  |  |  |  |  |  |  |  |  |  |
|  |  |  |  |  |  |  |  |  |  |  |  |  |  |
|  |  |  |  |  |  |  |  |  |  |  |  |  |  |
|  |  |  |  |  |  |  |  |  |  |  |  |  |  |
|  |  |  |  |  |  |  |  |  |  |  |  |  |  |
|  |  |  |  |  |  |  |  |  |  |  |  |  |  |
|  |  |  |  |  |  |  |  |  |  |  |  |  |  |
|  |  |  |  |  |  |  |  |  |  |  |  |  |  |

## REMARKS & HAPPENINGS

# SHIP'S LOG

DATE_____CREW_____GUESTS_____PAGE_____

DEPARTURE TIME_____DESTINATION_____ARRIVAL TIME_____

| TIME | ENGINE HOURS | PILOT INT'L | POSITION | HEADING | RPM | SPEED | ENGINE TEMP | AMPS | OIL | FUEL TANK | BARO | TEMP | WIND |
|------|------|------|------|------|------|------|------|------|------|------|------|------|------|
| | | | | | | | | | | | | | |
| | | | | | | | | | | | | | |
| | | | | | | | | | | | | | |
| | | | | | | | | | | | | | |
| | | | | | | | | | | | | | |
| | | | | | | | | | | | | | |
| | | | | | | | | | | | | | |
| | | | | | | | | | | | | | |
| | | | | | | | | | | | | | |
| | | | | | | | | | | | | | |
| | | | | | | | | | | | | | |
| | | | | | | | | | | | | | |
| | | | | | | | | | | | | | |
| | | | | | | | | | | | | | |
| | | | | | | | | | | | | | |
| | | | | | | | | | | | | | |

## REMARKS & HAPPENINGS

# SHIP'S LOG

DATE_____CREW_____GUESTS_____PAGE_____

DEPARTURE TIME_____DESTINATION_____ARRIVAL TIME_____

| TIME | ENGINE HOURS | PILOT INT'L | POSITION | HEADING | RPM | SPEED | ENGINE TEMP | AMPS | OIL | FUEL TANK | BARO | TEMP | WIND |
|------|------|------|------|------|------|------|------|------|------|------|------|------|------|
|  |  |  |  |  |  |  |  |  |  |  |  |  |  |
|  |  |  |  |  |  |  |  |  |  |  |  |  |  |
|  |  |  |  |  |  |  |  |  |  |  |  |  |  |
|  |  |  |  |  |  |  |  |  |  |  |  |  |  |
|  |  |  |  |  |  |  |  |  |  |  |  |  |  |
|  |  |  |  |  |  |  |  |  |  |  |  |  |  |
|  |  |  |  |  |  |  |  |  |  |  |  |  |  |
|  |  |  |  |  |  |  |  |  |  |  |  |  |  |
|  |  |  |  |  |  |  |  |  |  |  |  |  |  |
|  |  |  |  |  |  |  |  |  |  |  |  |  |  |
|  |  |  |  |  |  |  |  |  |  |  |  |  |  |
|  |  |  |  |  |  |  |  |  |  |  |  |  |  |
|  |  |  |  |  |  |  |  |  |  |  |  |  |  |
|  |  |  |  |  |  |  |  |  |  |  |  |  |  |

## REMARKS & HAPPENINGS

# SHIP'S LOG

DATE_____CREW_____GUESTS_____PAGE_____

DEPARTURE TIME_____DESTINATION_____ARRIVAL TIME_____

| TIME | ENGINE HOURS | PILOT INT'L | POSITION | HEADING | RPM | SPEED | ENGINE TEMP | AMPS | OIL | FUEL TANK | BARO | TEMP | WIND |
|------|------|------|------|------|------|------|------|------|------|------|------|------|------|
| | | | | | | | | | | | | | |
| | | | | | | | | | | | | | |
| | | | | | | | | | | | | | |
| | | | | | | | | | | | | | |
| | | | | | | | | | | | | | |
| | | | | | | | | | | | | | |
| | | | | | | | | | | | | | |
| | | | | | | | | | | | | | |
| | | | | | | | | | | | | | |
| | | | | | | | | | | | | | |
| | | | | | | | | | | | | | |
| | | | | | | | | | | | | | |
| | | | | | | | | | | | | | |
| | | | | | | | | | | | | | |
| | | | | | | | | | | | | | |
| | | | | | | | | | | | | | |

## REMARKS & HAPPENINGS

# SHIP'S LOG

DATE_____CREW_____GUESTS_____PAGE_____

DEPARTURE TIME_____DESTINATION_____ARRIVAL TIME_____

| TIME | ENGINE HOURS | PILOT INT'L | POSITION | HEADING | RPM | SPEED | ENGINE TEMP | AMPS | OIL | FUEL TANK | BARO | TEMP | WIND |
|------|------|------|------|------|------|------|------|------|------|------|------|------|------|
| | | | | | | | | | | | | | |
| | | | | | | | | | | | | | |
| | | | | | | | | | | | | | |
| | | | | | | | | | | | | | |
| | | | | | | | | | | | | | |
| | | | | | | | | | | | | | |
| | | | | | | | | | | | | | |
| | | | | | | | | | | | | | |
| | | | | | | | | | | | | | |
| | | | | | | | | | | | | | |
| | | | | | | | | | | | | | |
| | | | | | | | | | | | | | |
| | | | | | | | | | | | | | |

## REMARKS & HAPPENINGS

# SHIP'S LOG

DATE_____CREW_____GUESTS_____PAGE_____

DEPARTURE TIME_____DESTINATION_____ARRIVAL TIME_____

| TIME | ENGINE HOURS | PILOT INT'L | POSITION | HEADING | RPM | SPEED | ENGINE TEMP | AMPS | OIL | FUEL TANK | BARO | TEMP | WIND |
|------|------|------|------|------|------|------|------|------|------|------|------|------|------|
|  |  |  |  |  |  |  |  |  |  |  |  |  |  |
|  |  |  |  |  |  |  |  |  |  |  |  |  |  |
|  |  |  |  |  |  |  |  |  |  |  |  |  |  |
|  |  |  |  |  |  |  |  |  |  |  |  |  |  |
|  |  |  |  |  |  |  |  |  |  |  |  |  |  |
|  |  |  |  |  |  |  |  |  |  |  |  |  |  |
|  |  |  |  |  |  |  |  |  |  |  |  |  |  |
|  |  |  |  |  |  |  |  |  |  |  |  |  |  |
|  |  |  |  |  |  |  |  |  |  |  |  |  |  |
|  |  |  |  |  |  |  |  |  |  |  |  |  |  |
|  |  |  |  |  |  |  |  |  |  |  |  |  |  |
|  |  |  |  |  |  |  |  |  |  |  |  |  |  |
|  |  |  |  |  |  |  |  |  |  |  |  |  |  |
|  |  |  |  |  |  |  |  |  |  |  |  |  |  |

## REMARKS & HAPPENINGS

# SHIP'S LOG

DATE_____CREW_____GUESTS_____PAGE_____

DEPARTURE TIME_____DESTINATION_____ARRIVAL TIME_____

| TIME | ENGINE HOURS | PILOT INT'L | POSITION | HEADING | RPM | SPEED | ENGINE TEMP | AMPS | OIL | FUEL TANK | BARO | TEMP | WIND |
|------|------|------|------|------|------|------|------|------|------|------|------|------|------|
|  |  |  |  |  |  |  |  |  |  |  |  |  |  |
|  |  |  |  |  |  |  |  |  |  |  |  |  |  |
|  |  |  |  |  |  |  |  |  |  |  |  |  |  |
|  |  |  |  |  |  |  |  |  |  |  |  |  |  |
|  |  |  |  |  |  |  |  |  |  |  |  |  |  |
|  |  |  |  |  |  |  |  |  |  |  |  |  |  |
|  |  |  |  |  |  |  |  |  |  |  |  |  |  |
|  |  |  |  |  |  |  |  |  |  |  |  |  |  |
|  |  |  |  |  |  |  |  |  |  |  |  |  |  |
|  |  |  |  |  |  |  |  |  |  |  |  |  |  |
|  |  |  |  |  |  |  |  |  |  |  |  |  |  |
|  |  |  |  |  |  |  |  |  |  |  |  |  |  |

## REMARKS & HAPPENINGS

# SHIP'S LOG

DATE_____CREW_____GUESTS_____PAGE_____

DEPARTURE TIME_____DESTINATION_____ARRIVAL TIME_____

| TIME | ENGINE HOURS | PILOT INT'L | POSITION | HEADING | RPM | SPEED | ENGINE TEMP | AMPS | OIL | FUEL TANK | BARO | TEMP | WIND |
|------|------|------|------|------|------|------|------|------|------|------|------|------|------|
| | | | | | | | | | | | | | |
| | | | | | | | | | | | | | |
| | | | | | | | | | | | | | |
| | | | | | | | | | | | | | |
| | | | | | | | | | | | | | |
| | | | | | | | | | | | | | |
| | | | | | | | | | | | | | |
| | | | | | | | | | | | | | |
| | | | | | | | | | | | | | |
| | | | | | | | | | | | | | |
| | | | | | | | | | | | | | |
| | | | | | | | | | | | | | |
| | | | | | | | | | | | | | |
| | | | | | | | | | | | | | |
| | | | | | | | | | | | | | |
| | | | | | | | | | | | | | |

## REMARKS & HAPPENINGS

# SHIP'S LOG

DATE_____CREW_____GUESTS_____PAGE_____

DEPARTURE TIME_____DESTINATION_____ARRIVAL TIME_____

| TIME | ENGINE HOURS | PILOT INT'L | POSITION | HEADING | RPM | SPEED | ENGINE TEMP | AMPS | OIL | FUEL TANK | BARO | TEMP | WIND |
|------|--------------|-------------|----------|---------|-----|-------|-------------|------|-----|-----------|------|------|------|
|      |              |             |          |         |     |       |             |      |     |           |      |      |      |
|      |              |             |          |         |     |       |             |      |     |           |      |      |      |
|      |              |             |          |         |     |       |             |      |     |           |      |      |      |
|      |              |             |          |         |     |       |             |      |     |           |      |      |      |
|      |              |             |          |         |     |       |             |      |     |           |      |      |      |
|      |              |             |          |         |     |       |             |      |     |           |      |      |      |
|      |              |             |          |         |     |       |             |      |     |           |      |      |      |
|      |              |             |          |         |     |       |             |      |     |           |      |      |      |
|      |              |             |          |         |     |       |             |      |     |           |      |      |      |
|      |              |             |          |         |     |       |             |      |     |           |      |      |      |
|      |              |             |          |         |     |       |             |      |     |           |      |      |      |
|      |              |             |          |         |     |       |             |      |     |           |      |      |      |

## REMARKS & HAPPENINGS

# SHIP'S LOG

DATE_____CREW_____GUESTS_____PAGE_____

DEPARTURE TIME_____DESTINATION_____ARRIVAL TIME_____

| TIME | ENGINE HOURS | PILOT INT'L | POSITION | HEADING | RPM | SPEED | ENGINE TEMP | AMPS | OIL | FUEL TANK | BARO | TEMP | WIND |
|------|--------------|-------------|----------|---------|-----|-------|-------------|------|-----|-----------|------|------|------|
|      |              |             |          |         |     |       |             |      |     |           |      |      |      |
|      |              |             |          |         |     |       |             |      |     |           |      |      |      |
|      |              |             |          |         |     |       |             |      |     |           |      |      |      |
|      |              |             |          |         |     |       |             |      |     |           |      |      |      |
|      |              |             |          |         |     |       |             |      |     |           |      |      |      |
|      |              |             |          |         |     |       |             |      |     |           |      |      |      |
|      |              |             |          |         |     |       |             |      |     |           |      |      |      |
|      |              |             |          |         |     |       |             |      |     |           |      |      |      |
|      |              |             |          |         |     |       |             |      |     |           |      |      |      |
|      |              |             |          |         |     |       |             |      |     |           |      |      |      |
|      |              |             |          |         |     |       |             |      |     |           |      |      |      |
|      |              |             |          |         |     |       |             |      |     |           |      |      |      |
|      |              |             |          |         |     |       |             |      |     |           |      |      |      |
|      |              |             |          |         |     |       |             |      |     |           |      |      |      |

## REMARKS & HAPPENINGS

# SHIP'S LOG

DATE_____CREW_____GUESTS_____PAGE_____

DEPARTURE TIME_____DESTINATION_____ARRIVAL TIME_____

| TIME | ENGINE HOURS | PILOT INT'L | POSITION | HEADING | RPM | SPEED | ENGINE TEMP | AMPS | OIL | FUEL TANK | BARO | TEMP | WIND |
|------|------|------|------|------|------|------|------|------|------|------|------|------|------|
|  |  |  |  |  |  |  |  |  |  |  |  |  |  |
|  |  |  |  |  |  |  |  |  |  |  |  |  |  |
|  |  |  |  |  |  |  |  |  |  |  |  |  |  |
|  |  |  |  |  |  |  |  |  |  |  |  |  |  |
|  |  |  |  |  |  |  |  |  |  |  |  |  |  |
|  |  |  |  |  |  |  |  |  |  |  |  |  |  |
|  |  |  |  |  |  |  |  |  |  |  |  |  |  |
|  |  |  |  |  |  |  |  |  |  |  |  |  |  |
|  |  |  |  |  |  |  |  |  |  |  |  |  |  |
|  |  |  |  |  |  |  |  |  |  |  |  |  |  |
|  |  |  |  |  |  |  |  |  |  |  |  |  |  |
|  |  |  |  |  |  |  |  |  |  |  |  |  |  |

## REMARKS & HAPPENINGS

# SHIP'S LOG

DATE_____CREW_____GUESTS_____PAGE_____

DEPARTURE TIME_____DESTINATION_____ARRIVAL TIME_____

| TIME | ENGINE HOURS | PILOT INT'L | POSITION | HEADING | RPM | SPEED | ENGINE TEMP | AMPS | OIL | FUEL TANK | BARO | TEMP | WIND |
|------|------|------|------|------|------|------|------|------|------|------|------|------|------|
| | | | | | | | | | | | | | |
| | | | | | | | | | | | | | |
| | | | | | | | | | | | | | |
| | | | | | | | | | | | | | |
| | | | | | | | | | | | | | |
| | | | | | | | | | | | | | |
| | | | | | | | | | | | | | |
| | | | | | | | | | | | | | |
| | | | | | | | | | | | | | |
| | | | | | | | | | | | | | |
| | | | | | | | | | | | | | |
| | | | | | | | | | | | | | |
| | | | | | | | | | | | | | |
| | | | | | | | | | | | | | |
| | | | | | | | | | | | | | |
| | | | | | | | | | | | | | |

## REMARKS & HAPPENINGS

# SHIP'S LOG

DATE_____CREW_____GUESTS_____PAGE_____

DEPARTURE TIME_____DESTINATION_____ARRIVAL TIME_____

| TIME | ENGINE HOURS | PILOT INT'L | POSITION | HEADING | RPM | SPEED | ENGINE TEMP | AMPS | OIL | FUEL TANK | BARO | TEMP | WIND |
|------|------|------|------|------|------|------|------|------|------|------|------|------|------|
|  |  |  |  |  |  |  |  |  |  |  |  |  |  |
|  |  |  |  |  |  |  |  |  |  |  |  |  |  |
|  |  |  |  |  |  |  |  |  |  |  |  |  |  |
|  |  |  |  |  |  |  |  |  |  |  |  |  |  |
|  |  |  |  |  |  |  |  |  |  |  |  |  |  |
|  |  |  |  |  |  |  |  |  |  |  |  |  |  |
|  |  |  |  |  |  |  |  |  |  |  |  |  |  |
|  |  |  |  |  |  |  |  |  |  |  |  |  |  |
|  |  |  |  |  |  |  |  |  |  |  |  |  |  |
|  |  |  |  |  |  |  |  |  |  |  |  |  |  |
|  |  |  |  |  |  |  |  |  |  |  |  |  |  |
|  |  |  |  |  |  |  |  |  |  |  |  |  |  |

## REMARKS & HAPPENINGS

# SHIP'S LOG

DATE_____CREW_____GUESTS_____PAGE_____

DEPARTURE TIME_____DESTINATION_____ARRIVAL TIME_____

| TIME | ENGINE HOURS | PILOT INT'L | POSITION | HEADING | RPM | SPEED | ENGINE TEMP | AMPS | OIL | FUEL TANK | BARO | TEMP | WIND |
|------|------|------|------|------|------|------|------|------|------|------|------|------|------|
| | | | | | | | | | | | | | |
| | | | | | | | | | | | | | |
| | | | | | | | | | | | | | |
| | | | | | | | | | | | | | |
| | | | | | | | | | | | | | |
| | | | | | | | | | | | | | |
| | | | | | | | | | | | | | |
| | | | | | | | | | | | | | |
| | | | | | | | | | | | | | |
| | | | | | | | | | | | | | |
| | | | | | | | | | | | | | |
| | | | | | | | | | | | | | |
| | | | | | | | | | | | | | |
| | | | | | | | | | | | | | |
| | | | | | | | | | | | | | |
| | | | | | | | | | | | | | |

## REMARKS & HAPPENINGS

# SHIP'S LOG

DATE_____CREW_____GUESTS_____PAGE_____

DEPARTURE TIME_____DESTINATION_____ARRIVAL TIME_____

| TIME | ENGINE HOURS | PILOT INT'L | POSITION | HEADING | RPM | SPEED | ENGINE TEMP | AMPS | OIL | FUEL TANK | BARO | TEMP | WIND |
|------|------|------|------|------|------|------|------|------|------|------|------|------|------|
| | | | | | | | | | | | | | |
| | | | | | | | | | | | | | |
| | | | | | | | | | | | | | |
| | | | | | | | | | | | | | |
| | | | | | | | | | | | | | |
| | | | | | | | | | | | | | |
| | | | | | | | | | | | | | |
| | | | | | | | | | | | | | |
| | | | | | | | | | | | | | |
| | | | | | | | | | | | | | |
| | | | | | | | | | | | | | |
| | | | | | | | | | | | | | |
| | | | | | | | | | | | | | |
| | | | | | | | | | | | | | |
| | | | | | | | | | | | | | |

## REMARKS & HAPPENINGS

# SHIP'S LOG

DATE_____CREW_____GUESTS_____PAGE_____

DEPARTURE TIME_____DESTINATION_____ARRIVAL TIME_____

| TIME | ENGINE HOURS | PILOT INT'L | POSITION | HEADING | RPM | SPEED | ENGINE TEMP | AMPS | OIL | FUEL TANK | BARO | TEMP | WIND |
|------|------|------|------|------|------|------|------|------|------|------|------|------|------|
|  |  |  |  |  |  |  |  |  |  |  |  |  |  |
|  |  |  |  |  |  |  |  |  |  |  |  |  |  |
|  |  |  |  |  |  |  |  |  |  |  |  |  |  |
|  |  |  |  |  |  |  |  |  |  |  |  |  |  |
|  |  |  |  |  |  |  |  |  |  |  |  |  |  |
|  |  |  |  |  |  |  |  |  |  |  |  |  |  |
|  |  |  |  |  |  |  |  |  |  |  |  |  |  |
|  |  |  |  |  |  |  |  |  |  |  |  |  |  |
|  |  |  |  |  |  |  |  |  |  |  |  |  |  |
|  |  |  |  |  |  |  |  |  |  |  |  |  |  |
|  |  |  |  |  |  |  |  |  |  |  |  |  |  |
|  |  |  |  |  |  |  |  |  |  |  |  |  |  |

## REMARKS & HAPPENINGS

# SHIP'S LOG

DATE_____CREW_____GUESTS_____PAGE_____

DEPARTURE TIME_____DESTINATION_____ARRIVAL TIME_____

| TIME | ENGINE HOURS | PILOT INT'L | POSITION | HEADING | RPM | SPEED | ENGINE TEMP | AMPS | OIL | FUEL TANK | BARO | TEMP | WIND |
|------|------|------|------|------|------|------|------|------|------|------|------|------|------|
|  |  |  |  |  |  |  |  |  |  |  |  |  |  |
|  |  |  |  |  |  |  |  |  |  |  |  |  |  |
|  |  |  |  |  |  |  |  |  |  |  |  |  |  |
|  |  |  |  |  |  |  |  |  |  |  |  |  |  |
|  |  |  |  |  |  |  |  |  |  |  |  |  |  |
|  |  |  |  |  |  |  |  |  |  |  |  |  |  |
|  |  |  |  |  |  |  |  |  |  |  |  |  |  |
|  |  |  |  |  |  |  |  |  |  |  |  |  |  |
|  |  |  |  |  |  |  |  |  |  |  |  |  |  |
|  |  |  |  |  |  |  |  |  |  |  |  |  |  |
|  |  |  |  |  |  |  |  |  |  |  |  |  |  |
|  |  |  |  |  |  |  |  |  |  |  |  |  |  |
|  |  |  |  |  |  |  |  |  |  |  |  |  |  |

## REMARKS & HAPPENINGS

# SHIP'S LOG

DATE_____CREW_____GUESTS_____PAGE_____

DEPARTURE TIME_____DESTINATION_____ARRIVAL TIME_____

| TIME | ENGINE HOURS | PILOT INT'L | POSITION | HEADING | RPM | SPEED | ENGINE TEMP | AMPS | OIL | FUEL TANK | BARO | TEMP | WIND |
|------|------|------|------|------|------|------|------|------|------|------|------|------|------|
| | | | | | | | | | | | | | |
| | | | | | | | | | | | | | |
| | | | | | | | | | | | | | |
| | | | | | | | | | | | | | |
| | | | | | | | | | | | | | |
| | | | | | | | | | | | | | |
| | | | | | | | | | | | | | |
| | | | | | | | | | | | | | |
| | | | | | | | | | | | | | |
| | | | | | | | | | | | | | |
| | | | | | | | | | | | | | |
| | | | | | | | | | | | | | |
| | | | | | | | | | | | | | |
| | | | | | | | | | | | | | |
| | | | | | | | | | | | | | |
| | | | | | | | | | | | | | |
| | | | | | | | | | | | | | |
| | | | | | | | | | | | | | |
| | | | | | | | | | | | | | |

## REMARKS & HAPPENINGS

# SHIP'S LOG

DATE_____CREW_____GUESTS_____PAGE_____

DEPARTURE TIME_____DESTINATION_____ARRIVAL TIME_____

| TIME | ENGINE HOURS | PILOT INT'L | POSITION | HEADING | RPM | SPEED | ENGINE TEMP | AMPS | OIL | FUEL TANK | BARO | TEMP | WIND |
|---|---|---|---|---|---|---|---|---|---|---|---|---|---|
| | | | | | | | | | | | | | |
| | | | | | | | | | | | | | |
| | | | | | | | | | | | | | |
| | | | | | | | | | | | | | |
| | | | | | | | | | | | | | |
| | | | | | | | | | | | | | |
| | | | | | | | | | | | | | |
| | | | | | | | | | | | | | |
| | | | | | | | | | | | | | |
| | | | | | | | | | | | | | |
| | | | | | | | | | | | | | |
| | | | | | | | | | | | | | |
| | | | | | | | | | | | | | |
| | | | | | | | | | | | | | |
| | | | | | | | | | | | | | |

## REMARKS & HAPPENINGS

# SHIP'S LOG

DATE_____CREW_____GUESTS_____PAGE_____

DEPARTURE TIME_____DESTINATION_____ARRIVAL TIME_____

| TIME | ENGINE HOURS | PILOT INT'L | POSITION | HEADING | RPM | SPEED | ENGINE TEMP | AMPS | OIL | FUEL TANK | BARO | TEMP | WIND |
|------|------|------|------|------|------|------|------|------|------|------|------|------|------|
|  |  |  |  |  |  |  |  |  |  |  |  |  |  |
|  |  |  |  |  |  |  |  |  |  |  |  |  |  |
|  |  |  |  |  |  |  |  |  |  |  |  |  |  |
|  |  |  |  |  |  |  |  |  |  |  |  |  |  |
|  |  |  |  |  |  |  |  |  |  |  |  |  |  |
|  |  |  |  |  |  |  |  |  |  |  |  |  |  |
|  |  |  |  |  |  |  |  |  |  |  |  |  |  |
|  |  |  |  |  |  |  |  |  |  |  |  |  |  |
|  |  |  |  |  |  |  |  |  |  |  |  |  |  |
|  |  |  |  |  |  |  |  |  |  |  |  |  |  |
|  |  |  |  |  |  |  |  |  |  |  |  |  |  |
|  |  |  |  |  |  |  |  |  |  |  |  |  |  |
|  |  |  |  |  |  |  |  |  |  |  |  |  |  |

## REMARKS & HAPPENINGS

# SHIP'S LOG

DATE_____CREW_____GUESTS_____PAGE_____

DEPARTURE TIME_____DESTINATION_____ARRIVAL TIME_____

| TIME | ENGINE HOURS | PILOT INT'L | POSITION | HEADING | RPM | SPEED | ENGINE TEMP | AMPS | OIL | FUEL TANK | BARO | TEMP | WIND |
|------|------|------|------|------|------|------|------|------|------|------|------|------|------|
| | | | | | | | | | | | | | |
| | | | | | | | | | | | | | |
| | | | | | | | | | | | | | |
| | | | | | | | | | | | | | |
| | | | | | | | | | | | | | |
| | | | | | | | | | | | | | |
| | | | | | | | | | | | | | |
| | | | | | | | | | | | | | |
| | | | | | | | | | | | | | |
| | | | | | | | | | | | | | |
| | | | | | | | | | | | | | |
| | | | | | | | | | | | | | |
| | | | | | | | | | | | | | |
| | | | | | | | | | | | | | |
| | | | | | | | | | | | | | |
| | | | | | | | | | | | | | |

## REMARKS & HAPPENINGS

# SHIP'S LOG

DATE_____CREW_____GUESTS_____PAGE_____

DEPARTURE TIME_____DESTINATION_____ARRIVAL TIME_____

| TIME | ENGINE HOURS | PILOT INT'L | POSITION | HEADING | RPM | SPEED | ENGINE TEMP | AMPS | OIL | FUEL TANK | BARO | TEMP | WIND |
|------|------|------|------|------|------|------|------|------|------|------|------|------|------|
| | | | | | | | | | | | | | |
| | | | | | | | | | | | | | |
| | | | | | | | | | | | | | |
| | | | | | | | | | | | | | |
| | | | | | | | | | | | | | |
| | | | | | | | | | | | | | |
| | | | | | | | | | | | | | |
| | | | | | | | | | | | | | |
| | | | | | | | | | | | | | |
| | | | | | | | | | | | | | |
| | | | | | | | | | | | | | |
| | | | | | | | | | | | | | |
| | | | | | | | | | | | | | |
| | | | | | | | | | | | | | |
| | | | | | | | | | | | | | |
| | | | | | | | | | | | | | |

## REMARKS & HAPPENINGS

# SHIP'S LOG

DATE_____CREW_____GUESTS_____PAGE_____

DEPARTURE TIME_____DESTINATION_____ARRIVAL TIME_____

| TIME | ENGINE HOURS | PILOT INT'L | POSITION | HEADING | RPM | SPEED | ENGINE TEMP | AMPS | OIL | FUEL TANK | BARO | TEMP | WIND |
|------|------|------|------|------|------|------|------|------|------|------|------|------|------|
|  |  |  |  |  |  |  |  |  |  |  |  |  |  |
|  |  |  |  |  |  |  |  |  |  |  |  |  |  |
|  |  |  |  |  |  |  |  |  |  |  |  |  |  |
|  |  |  |  |  |  |  |  |  |  |  |  |  |  |
|  |  |  |  |  |  |  |  |  |  |  |  |  |  |
|  |  |  |  |  |  |  |  |  |  |  |  |  |  |
|  |  |  |  |  |  |  |  |  |  |  |  |  |  |
|  |  |  |  |  |  |  |  |  |  |  |  |  |  |
|  |  |  |  |  |  |  |  |  |  |  |  |  |  |
|  |  |  |  |  |  |  |  |  |  |  |  |  |  |
|  |  |  |  |  |  |  |  |  |  |  |  |  |  |
|  |  |  |  |  |  |  |  |  |  |  |  |  |  |
|  |  |  |  |  |  |  |  |  |  |  |  |  |  |
|  |  |  |  |  |  |  |  |  |  |  |  |  |  |

## REMARKS & HAPPENINGS

# SHIP'S LOG

DATE_____CREW_____GUESTS_____PAGE_____

DEPARTURE TIME_____DESTINATION_____ARRIVAL TIME_____

| TIME | ENGINE HOURS | PILOT INT'L | POSITION | HEADING | RPM | SPEED | ENGINE TEMP | AMPS | OIL | FUEL TANK | BARO | TEMP | WIND |
|------|------|------|------|------|------|------|------|------|------|------|------|------|------|
| | | | | | | | | | | | | | |
| | | | | | | | | | | | | | |
| | | | | | | | | | | | | | |
| | | | | | | | | | | | | | |
| | | | | | | | | | | | | | |
| | | | | | | | | | | | | | |
| | | | | | | | | | | | | | |
| | | | | | | | | | | | | | |
| | | | | | | | | | | | | | |
| | | | | | | | | | | | | | |
| | | | | | | | | | | | | | |
| | | | | | | | | | | | | | |
| | | | | | | | | | | | | | |
| | | | | | | | | | | | | | |
| | | | | | | | | | | | | | |
| | | | | | | | | | | | | | |
| | | | | | | | | | | | | | |

## REMARKS & HAPPENINGS

# SHIP'S LOG

DATE_____CREW_____GUESTS_____PAGE_____

DEPARTURE TIME_____DESTINATION_____ARRIVAL TIME_____

| TIME | ENGINE HOURS | PILOT INT'L | POSITION | HEADING | RPM | SPEED | ENGINE TEMP | AMPS | OIL | FUEL TANK | BARO | TEMP | WIND |
|------|--------------|-------------|----------|---------|-----|-------|-------------|------|-----|-----------|------|------|------|
|      |              |             |          |         |     |       |             |      |     |           |      |      |      |
|      |              |             |          |         |     |       |             |      |     |           |      |      |      |
|      |              |             |          |         |     |       |             |      |     |           |      |      |      |
|      |              |             |          |         |     |       |             |      |     |           |      |      |      |
|      |              |             |          |         |     |       |             |      |     |           |      |      |      |
|      |              |             |          |         |     |       |             |      |     |           |      |      |      |
|      |              |             |          |         |     |       |             |      |     |           |      |      |      |
|      |              |             |          |         |     |       |             |      |     |           |      |      |      |
|      |              |             |          |         |     |       |             |      |     |           |      |      |      |
|      |              |             |          |         |     |       |             |      |     |           |      |      |      |
|      |              |             |          |         |     |       |             |      |     |           |      |      |      |
|      |              |             |          |         |     |       |             |      |     |           |      |      |      |

## REMARKS & HAPPENINGS

# SHIP'S LOG

DATE_____CREW_____GUESTS_____PAGE_____

DEPARTURE TIME_____DESTINATION_____ARRIVAL TIME_____

| TIME | ENGINE HOURS | PILOT INT'L | POSITION | HEADING | RPM | SPEED | ENGINE TEMP | AMPS | OIL | FUEL TANK | BARO | TEMP | WIND |
|------|------|------|------|------|------|------|------|------|------|------|------|------|------|
| | | | | | | | | | | | | | |
| | | | | | | | | | | | | | |
| | | | | | | | | | | | | | |
| | | | | | | | | | | | | | |
| | | | | | | | | | | | | | |
| | | | | | | | | | | | | | |
| | | | | | | | | | | | | | |
| | | | | | | | | | | | | | |
| | | | | | | | | | | | | | |
| | | | | | | | | | | | | | |
| | | | | | | | | | | | | | |
| | | | | | | | | | | | | | |
| | | | | | | | | | | | | | |
| | | | | | | | | | | | | | |
| | | | | | | | | | | | | | |
| | | | | | | | | | | | | | |

## REMARKS & HAPPENINGS

# SHIP'S LOG

DATE_____CREW_____GUESTS_____PAGE_____

DEPARTURE TIME_____DESTINATION_____ARRIVAL TIME_____

| TIME | ENGINE HOURS | PILOT INT'L | POSITION | HEADING | RPM | SPEED | ENGINE TEMP | AMPS | OIL | FUEL TANK | BARO | TEMP | WIND |
|------|--------------|-------------|----------|---------|-----|-------|-------------|------|-----|-----------|------|------|------|
|      |              |             |          |         |     |       |             |      |     |           |      |      |      |
|      |              |             |          |         |     |       |             |      |     |           |      |      |      |
|      |              |             |          |         |     |       |             |      |     |           |      |      |      |
|      |              |             |          |         |     |       |             |      |     |           |      |      |      |
|      |              |             |          |         |     |       |             |      |     |           |      |      |      |
|      |              |             |          |         |     |       |             |      |     |           |      |      |      |
|      |              |             |          |         |     |       |             |      |     |           |      |      |      |
|      |              |             |          |         |     |       |             |      |     |           |      |      |      |
|      |              |             |          |         |     |       |             |      |     |           |      |      |      |
|      |              |             |          |         |     |       |             |      |     |           |      |      |      |
|      |              |             |          |         |     |       |             |      |     |           |      |      |      |
|      |              |             |          |         |     |       |             |      |     |           |      |      |      |
|      |              |             |          |         |     |       |             |      |     |           |      |      |      |
|      |              |             |          |         |     |       |             |      |     |           |      |      |      |

## REMARKS & HAPPENINGS

# SHIP'S LOG

DATE_____CREW_____GUESTS_____PAGE_____

DEPARTURE TIME_____DESTINATION_____ARRIVAL TIME_____

| TIME | ENGINE HOURS | PILOT INT'L | POSITION | HEADING | RPM | SPEED | ENGINE TEMP | AMPS | OIL | FUEL TANK | BARO | TEMP | WIND |
|------|------|------|------|------|------|------|------|------|------|------|------|------|------|
| | | | | | | | | | | | | | |
| | | | | | | | | | | | | | |
| | | | | | | | | | | | | | |
| | | | | | | | | | | | | | |
| | | | | | | | | | | | | | |
| | | | | | | | | | | | | | |
| | | | | | | | | | | | | | |
| | | | | | | | | | | | | | |
| | | | | | | | | | | | | | |
| | | | | | | | | | | | | | |
| | | | | | | | | | | | | | |
| | | | | | | | | | | | | | |
| | | | | | | | | | | | | | |
| | | | | | | | | | | | | | |
| | | | | | | | | | | | | | |
| | | | | | | | | | | | | | |

## REMARKS & HAPPENINGS

# SHIP'S LOG

DATE_____CREW_____GUESTS_____PAGE_____

DEPARTURE TIME_____DESTINATION_____ARRIVAL TIME_____

| TIME | ENGINE HOURS | PILOT INT'L | POSITION | HEADING | RPM | SPEED | ENGINE TEMP | AMPS | OIL | FUEL TANK | BARO | TEMP | WIND |
|------|------|------|------|------|------|------|------|------|------|------|------|------|------|
|  |  |  |  |  |  |  |  |  |  |  |  |  |  |
|  |  |  |  |  |  |  |  |  |  |  |  |  |  |
|  |  |  |  |  |  |  |  |  |  |  |  |  |  |
|  |  |  |  |  |  |  |  |  |  |  |  |  |  |
|  |  |  |  |  |  |  |  |  |  |  |  |  |  |
|  |  |  |  |  |  |  |  |  |  |  |  |  |  |
|  |  |  |  |  |  |  |  |  |  |  |  |  |  |
|  |  |  |  |  |  |  |  |  |  |  |  |  |  |
|  |  |  |  |  |  |  |  |  |  |  |  |  |  |
|  |  |  |  |  |  |  |  |  |  |  |  |  |  |
|  |  |  |  |  |  |  |  |  |  |  |  |  |  |
|  |  |  |  |  |  |  |  |  |  |  |  |  |  |
|  |  |  |  |  |  |  |  |  |  |  |  |  |  |

## REMARKS & HAPPENINGS

# SHIP'S LOG

DATE_____CREW_____GUESTS_____PAGE_____

DEPARTURE TIME_____DESTINATION_____ARRIVAL TIME_____

| TIME | ENGINE HOURS | PILOT INT'L | POSITION | HEADING | RPM | SPEED | ENGINE TEMP | AMPS | OIL | FUEL TANK | BARO | TEMP | WIND |
|------|--------------|-------------|----------|---------|-----|-------|-------------|------|-----|-----------|------|------|------|
|      |              |             |          |         |     |       |             |      |     |           |      |      |      |
|      |              |             |          |         |     |       |             |      |     |           |      |      |      |
|      |              |             |          |         |     |       |             |      |     |           |      |      |      |
|      |              |             |          |         |     |       |             |      |     |           |      |      |      |
|      |              |             |          |         |     |       |             |      |     |           |      |      |      |
|      |              |             |          |         |     |       |             |      |     |           |      |      |      |
|      |              |             |          |         |     |       |             |      |     |           |      |      |      |
|      |              |             |          |         |     |       |             |      |     |           |      |      |      |
|      |              |             |          |         |     |       |             |      |     |           |      |      |      |
|      |              |             |          |         |     |       |             |      |     |           |      |      |      |
|      |              |             |          |         |     |       |             |      |     |           |      |      |      |
|      |              |             |          |         |     |       |             |      |     |           |      |      |      |
|      |              |             |          |         |     |       |             |      |     |           |      |      |      |
|      |              |             |          |         |     |       |             |      |     |           |      |      |      |

## REMARKS & HAPPENINGS

# SHIP'S LOG

DATE_____CREW_____GUESTS_____PAGE_____

DEPARTURE TIME_____DESTINATION_____ARRIVAL TIME_____

| TIME | ENGINE HOURS | PILOT INT'L | POSITION | HEADING | RPM | SPEED | ENGINE TEMP | AMPS | OIL | FUEL TANK | BARO | TEMP | WIND |
|------|------|------|------|------|------|------|------|------|------|------|------|------|------|
|  |  |  |  |  |  |  |  |  |  |  |  |  |  |
|  |  |  |  |  |  |  |  |  |  |  |  |  |  |
|  |  |  |  |  |  |  |  |  |  |  |  |  |  |
|  |  |  |  |  |  |  |  |  |  |  |  |  |  |
|  |  |  |  |  |  |  |  |  |  |  |  |  |  |
|  |  |  |  |  |  |  |  |  |  |  |  |  |  |
|  |  |  |  |  |  |  |  |  |  |  |  |  |  |
|  |  |  |  |  |  |  |  |  |  |  |  |  |  |
|  |  |  |  |  |  |  |  |  |  |  |  |  |  |
|  |  |  |  |  |  |  |  |  |  |  |  |  |  |
|  |  |  |  |  |  |  |  |  |  |  |  |  |  |
|  |  |  |  |  |  |  |  |  |  |  |  |  |  |
|  |  |  |  |  |  |  |  |  |  |  |  |  |  |

## REMARKS & HAPPENINGS

# SHIP'S LOG

DATE_____CREW_____GUESTS_____PAGE_____

DEPARTURE TIME_____DESTINATION_____ARRIVAL TIME_____

| TIME | ENGINE HOURS | PILOT INT'L | POSITION | HEADING | RPM | SPEED | ENGINE TEMP | AMPS | OIL | FUEL TANK | BARO | TEMP | WIND |
|------|------|------|------|------|------|------|------|------|------|------|------|------|------|
|  |  |  |  |  |  |  |  |  |  |  |  |  |  |
|  |  |  |  |  |  |  |  |  |  |  |  |  |  |
|  |  |  |  |  |  |  |  |  |  |  |  |  |  |
|  |  |  |  |  |  |  |  |  |  |  |  |  |  |
|  |  |  |  |  |  |  |  |  |  |  |  |  |  |
|  |  |  |  |  |  |  |  |  |  |  |  |  |  |
|  |  |  |  |  |  |  |  |  |  |  |  |  |  |
|  |  |  |  |  |  |  |  |  |  |  |  |  |  |
|  |  |  |  |  |  |  |  |  |  |  |  |  |  |
|  |  |  |  |  |  |  |  |  |  |  |  |  |  |
|  |  |  |  |  |  |  |  |  |  |  |  |  |  |
|  |  |  |  |  |  |  |  |  |  |  |  |  |  |
|  |  |  |  |  |  |  |  |  |  |  |  |  |  |
|  |  |  |  |  |  |  |  |  |  |  |  |  |  |

## REMARKS & HAPPENINGS

# SHIP'S LOG

DATE_____CREW_____GUESTS_____PAGE____

DEPARTURE TIME_____DESTINATION_____ARRIVAL TIME_____

| TIME | ENGINE HOURS | PILOT INT'L | POSITION | HEADING | RPM | SPEED | ENGINE TEMP | AMPS | OIL | FUEL TANK | BARO | TEMP | WIND |
|------|------|------|------|------|------|------|------|------|------|------|------|------|------|
| | | | | | | | | | | | | | |
| | | | | | | | | | | | | | |
| | | | | | | | | | | | | | |
| | | | | | | | | | | | | | |
| | | | | | | | | | | | | | |
| | | | | | | | | | | | | | |
| | | | | | | | | | | | | | |
| | | | | | | | | | | | | | |
| | | | | | | | | | | | | | |
| | | | | | | | | | | | | | |
| | | | | | | | | | | | | | |
| | | | | | | | | | | | | | |
| | | | | | | | | | | | | | |
| | | | | | | | | | | | | | |
| | | | | | | | | | | | | | |
| | | | | | | | | | | | | | |

## REMARKS & HAPPENINGS

# SHIP'S LOG

DATE_____CREW_____GUESTS_____PAGE_____

DEPARTURE TIME_____DESTINATION_____ARRIVAL TIME_____

| TIME | ENGINE HOURS | PILOT INT'L | POSITION | HEADING | RPM | SPEED | ENGINE TEMP | AMPS | OIL | FUEL TANK | BARO | TEMP | WIND |
|------|------|------|------|------|------|------|------|------|------|------|------|------|------|
| | | | | | | | | | | | | | |
| | | | | | | | | | | | | | |
| | | | | | | | | | | | | | |
| | | | | | | | | | | | | | |
| | | | | | | | | | | | | | |
| | | | | | | | | | | | | | |
| | | | | | | | | | | | | | |
| | | | | | | | | | | | | | |
| | | | | | | | | | | | | | |
| | | | | | | | | | | | | | |
| | | | | | | | | | | | | | |
| | | | | | | | | | | | | | |
| | | | | | | | | | | | | | |
| | | | | | | | | | | | | | |
| | | | | | | | | | | | | | |
| | | | | | | | | | | | | | |

## REMARKS & HAPPENINGS

# SHIP'S LOG

DATE_____CREW_____GUESTS_____PAGE_____

DEPARTURE TIME_____DESTINATION_____ARRIVAL TIME_____

| TIME | ENGINE HOURS | PILOT INT'L | POSITION | HEADING | RPM | SPEED | ENGINE TEMP | AMPS | OIL | FUEL TANK | BARO | TEMP | WIND |
|------|--------------|-------------|----------|---------|-----|-------|-------------|------|-----|-----------|------|------|------|
|      |              |             |          |         |     |       |             |      |     |           |      |      |      |
|      |              |             |          |         |     |       |             |      |     |           |      |      |      |
|      |              |             |          |         |     |       |             |      |     |           |      |      |      |
|      |              |             |          |         |     |       |             |      |     |           |      |      |      |
|      |              |             |          |         |     |       |             |      |     |           |      |      |      |
|      |              |             |          |         |     |       |             |      |     |           |      |      |      |
|      |              |             |          |         |     |       |             |      |     |           |      |      |      |
|      |              |             |          |         |     |       |             |      |     |           |      |      |      |
|      |              |             |          |         |     |       |             |      |     |           |      |      |      |
|      |              |             |          |         |     |       |             |      |     |           |      |      |      |
|      |              |             |          |         |     |       |             |      |     |           |      |      |      |
|      |              |             |          |         |     |       |             |      |     |           |      |      |      |
|      |              |             |          |         |     |       |             |      |     |           |      |      |      |
|      |              |             |          |         |     |       |             |      |     |           |      |      |      |

## REMARKS & HAPPENINGS

# SHIP'S LOG

DATE_____CREW_____GUESTS_____PAGE_____

DEPARTURE TIME_____DESTINATION_____ARRIVAL TIME_____

| TIME | ENGINE HOURS | PILOT INT'L | POSITION | HEADING | RPM | SPEED | ENGINE TEMP | AMPS | OIL | FUEL TANK | BARO | TEMP | WIND |
|------|--------------|-------------|----------|---------|-----|-------|-------------|------|-----|-----------|------|------|------|
|      |              |             |          |         |     |       |             |      |     |           |      |      |      |
|      |              |             |          |         |     |       |             |      |     |           |      |      |      |
|      |              |             |          |         |     |       |             |      |     |           |      |      |      |
|      |              |             |          |         |     |       |             |      |     |           |      |      |      |
|      |              |             |          |         |     |       |             |      |     |           |      |      |      |
|      |              |             |          |         |     |       |             |      |     |           |      |      |      |
|      |              |             |          |         |     |       |             |      |     |           |      |      |      |
|      |              |             |          |         |     |       |             |      |     |           |      |      |      |
|      |              |             |          |         |     |       |             |      |     |           |      |      |      |
|      |              |             |          |         |     |       |             |      |     |           |      |      |      |
|      |              |             |          |         |     |       |             |      |     |           |      |      |      |
|      |              |             |          |         |     |       |             |      |     |           |      |      |      |
|      |              |             |          |         |     |       |             |      |     |           |      |      |      |
|      |              |             |          |         |     |       |             |      |     |           |      |      |      |
|      |              |             |          |         |     |       |             |      |     |           |      |      |      |

## REMARKS & HAPPENINGS

# SHIP'S LOG

DATE_____CREW_____GUESTS_____PAGE_____

DEPARTURE TIME_____DESTINATION_____ARRIVAL TIME_____

| TIME | ENGINE HOURS | PILOT INT'L | POSITION | HEADING | RPM | SPEED | ENGINE TEMP | AMPS | OIL | FUEL TANK | BARO | TEMP | WIND |
|------|------|------|------|------|------|------|------|------|------|------|------|------|------|
|  |  |  |  |  |  |  |  |  |  |  |  |  |  |
|  |  |  |  |  |  |  |  |  |  |  |  |  |  |
|  |  |  |  |  |  |  |  |  |  |  |  |  |  |
|  |  |  |  |  |  |  |  |  |  |  |  |  |  |
|  |  |  |  |  |  |  |  |  |  |  |  |  |  |
|  |  |  |  |  |  |  |  |  |  |  |  |  |  |
|  |  |  |  |  |  |  |  |  |  |  |  |  |  |
|  |  |  |  |  |  |  |  |  |  |  |  |  |  |
|  |  |  |  |  |  |  |  |  |  |  |  |  |  |
|  |  |  |  |  |  |  |  |  |  |  |  |  |  |
|  |  |  |  |  |  |  |  |  |  |  |  |  |  |
|  |  |  |  |  |  |  |  |  |  |  |  |  |  |

## REMARKS & HAPPENINGS

# SHIP'S LOG

DATE_____CREW_____GUESTS_____PAGE_____

DEPARTURE TIME_____DESTINATION_____ARRIVAL TIME_____

| TIME | ENGINE HOURS | PILOT INT'L | POSITION | HEADING | RPM | SPEED | ENGINE TEMP | AMPS | OIL | FUEL TANK | BARO | TEMP | WIND |
|------|------|------|------|------|------|------|------|------|------|------|------|------|------|
| | | | | | | | | | | | | | |
| | | | | | | | | | | | | | |
| | | | | | | | | | | | | | |
| | | | | | | | | | | | | | |
| | | | | | | | | | | | | | |
| | | | | | | | | | | | | | |
| | | | | | | | | | | | | | |
| | | | | | | | | | | | | | |
| | | | | | | | | | | | | | |
| | | | | | | | | | | | | | |
| | | | | | | | | | | | | | |
| | | | | | | | | | | | | | |
| | | | | | | | | | | | | | |
| | | | | | | | | | | | | | |
| | | | | | | | | | | | | | |
| | | | | | | | | | | | | | |
| | | | | | | | | | | | | | |

## REMARKS & HAPPENINGS

# FUEL LOG

| DATE | ADDITIVES | HOURS | VENDOR | GALS | PRICE/GAL | TOTAL COST | MILES/GAL | GAL/HOUR |
|------|-----------|-------|--------|------|-----------|------------|-----------|----------|
|      |           |       |        |      |           |            |           |          |
|      |           |       |        |      |           |            |           |          |
|      |           |       |        |      |           |            |           |          |
|      |           |       |        |      |           |            |           |          |
|      |           |       |        |      |           |            |           |          |
|      |           |       |        |      |           |            |           |          |
|      |           |       |        |      |           |            |           |          |
|      |           |       |        |      |           |            |           |          |
|      |           |       |        |      |           |            |           |          |
|      |           |       |        |      |           |            |           |          |
|      |           |       |        |      |           |            |           |          |
|      |           |       |        |      |           |            |           |          |
|      |           |       |        |      |           |            |           |          |
|      |           |       |        |      |           |            |           |          |
|      |           |       |        |      |           |            |           |          |
|      |           |       |        |      |           |            |           |          |
|      |           |       |        |      |           |            |           |          |
|      |           |       |        |      |           |            |           |          |
|      |           |       |        |      |           |            |           |          |
|      |           |       |        |      |           |            |           |          |
|      |           |       |        |      |           |            |           |          |
|      |           |       |        |      |           |            |           |          |
|      |           |       |        |      |           |            |           |          |
|      |           |       |        |      |           |            |           |          |
|      |           |       |        |      |           |            |           |          |
|      |           |       |        |      |           |            |           |          |
|      |           |       |        |      |           |            |           |          |
|      |           |       |        |      |           |            |           |          |
|      |           |       |        |      |           |            |           |          |
|      |           |       |        |      |           |            |           |          |

# FUEL LOG

| DATE | ADDITIVES | HOURS | VENDOR | GALS | PRICE/GAL | TOTAL COST | MILES/GAL | GAL/HOUR |
|------|-----------|-------|--------|------|-----------|------------|-----------|----------|
|      |           |       |        |      |           |            |           |          |
|      |           |       |        |      |           |            |           |          |
|      |           |       |        |      |           |            |           |          |
|      |           |       |        |      |           |            |           |          |
|      |           |       |        |      |           |            |           |          |
|      |           |       |        |      |           |            |           |          |
|      |           |       |        |      |           |            |           |          |
|      |           |       |        |      |           |            |           |          |
|      |           |       |        |      |           |            |           |          |
|      |           |       |        |      |           |            |           |          |
|      |           |       |        |      |           |            |           |          |
|      |           |       |        |      |           |            |           |          |
|      |           |       |        |      |           |            |           |          |
|      |           |       |        |      |           |            |           |          |
|      |           |       |        |      |           |            |           |          |
|      |           |       |        |      |           |            |           |          |
|      |           |       |        |      |           |            |           |          |
|      |           |       |        |      |           |            |           |          |
|      |           |       |        |      |           |            |           |          |
|      |           |       |        |      |           |            |           |          |
|      |           |       |        |      |           |            |           |          |
|      |           |       |        |      |           |            |           |          |
|      |           |       |        |      |           |            |           |          |
|      |           |       |        |      |           |            |           |          |
|      |           |       |        |      |           |            |           |          |
|      |           |       |        |      |           |            |           |          |
|      |           |       |        |      |           |            |           |          |
|      |           |       |        |      |           |            |           |          |
|      |           |       |        |      |           |            |           |          |
|      |           |       |        |      |           |            |           |          |

# FUEL LOG

| DATE | ADDITIVES | HOURS | VENDOR | GALS | PRICE/GAL | TOTAL COST | MILES/GAL | GAL/HOUR |
|------|-----------|-------|--------|------|-----------|------------|-----------|----------|
|      |           |       |        |      |           |            |           |          |
|      |           |       |        |      |           |            |           |          |
|      |           |       |        |      |           |            |           |          |
|      |           |       |        |      |           |            |           |          |
|      |           |       |        |      |           |            |           |          |
|      |           |       |        |      |           |            |           |          |
|      |           |       |        |      |           |            |           |          |
|      |           |       |        |      |           |            |           |          |
|      |           |       |        |      |           |            |           |          |
|      |           |       |        |      |           |            |           |          |
|      |           |       |        |      |           |            |           |          |
|      |           |       |        |      |           |            |           |          |
|      |           |       |        |      |           |            |           |          |
|      |           |       |        |      |           |            |           |          |
|      |           |       |        |      |           |            |           |          |
|      |           |       |        |      |           |            |           |          |
|      |           |       |        |      |           |            |           |          |
|      |           |       |        |      |           |            |           |          |
|      |           |       |        |      |           |            |           |          |
|      |           |       |        |      |           |            |           |          |
|      |           |       |        |      |           |            |           |          |
|      |           |       |        |      |           |            |           |          |
|      |           |       |        |      |           |            |           |          |
|      |           |       |        |      |           |            |           |          |
|      |           |       |        |      |           |            |           |          |
|      |           |       |        |      |           |            |           |          |
|      |           |       |        |      |           |            |           |          |
|      |           |       |        |      |           |            |           |          |
|      |           |       |        |      |           |            |           |          |
|      |           |       |        |      |           |            |           |          |

# FUEL LOG

| DATE | ADDITIVES | HOURS | VENDOR | GALS | PRICE/GAL | TOTAL COST | MILES/GAL | GAL/HOUR |
|------|-----------|-------|--------|------|-----------|------------|-----------|----------|
|      |           |       |        |      |           |            |           |          |
|      |           |       |        |      |           |            |           |          |
|      |           |       |        |      |           |            |           |          |
|      |           |       |        |      |           |            |           |          |
|      |           |       |        |      |           |            |           |          |
|      |           |       |        |      |           |            |           |          |
|      |           |       |        |      |           |            |           |          |
|      |           |       |        |      |           |            |           |          |
|      |           |       |        |      |           |            |           |          |
|      |           |       |        |      |           |            |           |          |
|      |           |       |        |      |           |            |           |          |
|      |           |       |        |      |           |            |           |          |
|      |           |       |        |      |           |            |           |          |
|      |           |       |        |      |           |            |           |          |
|      |           |       |        |      |           |            |           |          |
|      |           |       |        |      |           |            |           |          |
|      |           |       |        |      |           |            |           |          |
|      |           |       |        |      |           |            |           |          |
|      |           |       |        |      |           |            |           |          |
|      |           |       |        |      |           |            |           |          |
|      |           |       |        |      |           |            |           |          |
|      |           |       |        |      |           |            |           |          |
|      |           |       |        |      |           |            |           |          |
|      |           |       |        |      |           |            |           |          |

# WAYPOINT

| WPT | L & L | VISUAL AID | CHART NUMBER | AREA | DATE |
|-----|-------|-----------|--------------|------|------|
|     |       |           |              |      |      |
|     |       |           |              |      |      |
|     |       |           |              |      |      |
|     |       |           |              |      |      |
|     |       |           |              |      |      |
|     |       |           |              |      |      |
|     |       |           |              |      |      |
|     |       |           |              |      |      |
|     |       |           |              |      |      |
|     |       |           |              |      |      |
|     |       |           |              |      |      |
|     |       |           |              |      |      |
|     |       |           |              |      |      |
|     |       |           |              |      |      |
|     |       |           |              |      |      |
|     |       |           |              |      |      |
|     |       |           |              |      |      |
|     |       |           |              |      |      |
|     |       |           |              |      |      |
|     |       |           |              |      |      |
|     |       |           |              |      |      |
|     |       |           |              |      |      |
|     |       |           |              |      |      |
|     |       |           |              |      |      |
|     |       |           |              |      |      |
|     |       |           |              |      |      |
|     |       |           |              |      |      |
|     |       |           |              |      |      |
|     |       |           |              |      |      |
|     |       |           |              |      |      |

# WAYPOINT

| WPT | L & L | VISUAL AID | CHART NUMBER | AREA | DATE |
|---|---|---|---|---|---|
| | | | | | |
| | | | | | |
| | | | | | |
| | | | | | |
| | | | | | |
| | | | | | |
| | | | | | |
| | | | | | |
| | | | | | |
| | | | | | |
| | | | | | |
| | | | | | |
| | | | | | |
| | | | | | |
| | | | | | |
| | | | | | |
| | | | | | |
| | | | | | |
| | | | | | |
| | | | | | |
| | | | | | |
| | | | | | |
| | | | | | |
| | | | | | |
| | | | | | |
| | | | | | |
| | | | | | |
| | | | | | |
| | | | | | |
| | | | | | |
| | | | | | |
| | | | | | |
| | | | | | |
| | | | | | |
| | | | | | |
| | | | | | |
| | | | | | |
| | | | | | |
| | | | | | |
| | | | | | |
| | | | | | |
| | | | | | |
| | | | | | |
| | | | | | |
| | | | | | |
| | | | | | |

# GENSET RUN TIME

| DATE | HOURS | TIME ON | CHARGE READING | TIME OFF | RUN HOURS | CHARGE READING | TOTAL RUN TIME |
|------|-------|---------|----------------|----------|-----------|----------------|----------------|
|      |       |         |                |          |           |                |                |
|      |       |         |                |          |           |                |                |
|      |       |         |                |          |           |                |                |
|      |       |         |                |          |           |                |                |
|      |       |         |                |          |           |                |                |
|      |       |         |                |          |           |                |                |
|      |       |         |                |          |           |                |                |
|      |       |         |                |          |           |                |                |
|      |       |         |                |          |           |                |                |
|      |       |         |                |          |           |                |                |
|      |       |         |                |          |           |                |                |
|      |       |         |                |          |           |                |                |
|      |       |         |                |          |           |                |                |
|      |       |         |                |          |           |                |                |
|      |       |         |                |          |           |                |                |
|      |       |         |                |          |           |                |                |
|      |       |         |                |          |           |                |                |
|      |       |         |                |          |           |                |                |
|      |       |         |                |          |           |                |                |

# GENSET RUN TIME

| DATE | HOURS | TIME ON | CHARGE READING | TIME OFF | RUN HOURS | CHARGE READING | TOTAL RUN TIME |
|------|-------|---------|----------------|----------|-----------|----------------|----------------|
|      |       |         |                |          |           |                |                |
|      |       |         |                |          |           |                |                |
|      |       |         |                |          |           |                |                |
|      |       |         |                |          |           |                |                |
|      |       |         |                |          |           |                |                |
|      |       |         |                |          |           |                |                |
|      |       |         |                |          |           |                |                |
|      |       |         |                |          |           |                |                |
|      |       |         |                |          |           |                |                |
|      |       |         |                |          |           |                |                |
|      |       |         |                |          |           |                |                |
|      |       |         |                |          |           |                |                |
|      |       |         |                |          |           |                |                |
|      |       |         |                |          |           |                |                |
|      |       |         |                |          |           |                |                |
|      |       |         |                |          |           |                |                |
|      |       |         |                |          |           |                |                |
|      |       |         |                |          |           |                |                |
|      |       |         |                |          |           |                |                |
|      |       |         |                |          |           |                |                |
|      |       |         |                |          |           |                |                |
|      |       |         |                |          |           |                |                |
|      |       |         |                |          |           |                |                |
|      |       |         |                |          |           |                |                |
|      |       |         |                |          |           |                |                |
|      |       |         |                |          |           |                |                |
|      |       |         |                |          |           |                |                |
|      |       |         |                |          |           |                |                |
|      |       |         |                |          |           |                |                |
|      |       |         |                |          |           |                |                |

# GENSET RUN TIME

| DATE | HOURS | TIME ON | CHARGE READING | TIME OFF | RUN HOURS | CHARGE READING | TOTAL RUN TIME |
|------|-------|---------|----------------|----------|-----------|----------------|----------------|
|      |       |         |                |          |           |                |                |
|      |       |         |                |          |           |                |                |
|      |       |         |                |          |           |                |                |
|      |       |         |                |          |           |                |                |
|      |       |         |                |          |           |                |                |
|      |       |         |                |          |           |                |                |
|      |       |         |                |          |           |                |                |
|      |       |         |                |          |           |                |                |
|      |       |         |                |          |           |                |                |
|      |       |         |                |          |           |                |                |
|      |       |         |                |          |           |                |                |
|      |       |         |                |          |           |                |                |
|      |       |         |                |          |           |                |                |
|      |       |         |                |          |           |                |                |
|      |       |         |                |          |           |                |                |
|      |       |         |                |          |           |                |                |
|      |       |         |                |          |           |                |                |
|      |       |         |                |          |           |                |                |
|      |       |         |                |          |           |                |                |
|      |       |         |                |          |           |                |                |
|      |       |         |                |          |           |                |                |
|      |       |         |                |          |           |                |                |
|      |       |         |                |          |           |                |                |
|      |       |         |                |          |           |                |                |
|      |       |         |                |          |           |                |                |
|      |       |         |                |          |           |                |                |
|      |       |         |                |          |           |                |                |
|      |       |         |                |          |           |                |                |
|      |       |         |                |          |           |                |                |
|      |       |         |                |          |           |                |                |
|      |       |         |                |          |           |                |                |
|      |       |         |                |          |           |                |                |
|      |       |         |                |          |           |                |                |
|      |       |         |                |          |           |                |                |

# GENSET RUN TIME

| DATE | HOURS | TIME ON | CHARGE READING | TIME OFF | RUN HOURS | CHARGE READING | TOTAL RUN TIME |
|------|-------|---------|----------------|----------|-----------|----------------|----------------|
| | | | | | | | |
| | | | | | | | |
| | | | | | | | |
| | | | | | | | |
| | | | | | | | |
| | | | | | | | |
| | | | | | | | |
| | | | | | | | |
| | | | | | | | |
| | | | | | | | |
| | | | | | | | |
| | | | | | | | |
| | | | | | | | |
| | | | | | | | |
| | | | | | | | |
| | | | | | | | |
| | | | | | | | |
| | | | | | | | |
| | | | | | | | |
| | | | | | | | |
| | | | | | | | |
| | | | | | | | |
| | | | | | | | |
| | | | | | | | |
| | | | | | | | |
| | | | | | | | |
| | | | | | | | |
| | | | | | | | |
| | | | | | | | |
| | | | | | | | |
| | | | | | | | |
| | | | | | | | |
| | | | | | | | |
| | | | | | | | |
| | | | | | | | |
| | | | | | | | |

# ENGINE MAINTENANCE

| DATE | ENGINE | HOURS | OIL | TRANS | COOLANT | BATTERIES | | |
|------|--------|-------|-----|-------|---------|-----------|------|--------|
| | | | | | | ENGINE | HOUSE | GENSET |
| | | | | | | | | |
| | | | | | | | | |
| | | | | | | | | |
| | | | | | | | | |
| | | | | | | | | |
| | | | | | | | | |
| | | | | | | | | |
| | | | | | | | | |
| | | | | | | | | |
| | | | | | | | | |
| | | | | | | | | |
| | | | | | | | | |
| | | | | | | | | |
| | | | | | | | | |
| | | | | | | | | |
| | | | | | | | | |
| | | | | | | | | |
| | | | | | | | | |
| | | | | | | | | |
| | | | | | | | | |
| | | | | | | | | |
| | | | | | | | | |
| | | | | | | | | |
| | | | | | | | | |
| | | | | | | | | |
| | | | | | | | | |
| | | | | | | | | |
| | | | | | | | | |
| | | | | | | | | |
| | | | | | | | | |
| | | | | | | | | |
| | | | | | | | | |
| | | | | | | | | |

# ENGINE MAINTENANCE

| DATE | ENGINE | HOURS | OIL | TRANS | COOLANT | BATTERIES | | |
|------|--------|-------|-----|-------|---------|-----------|-----|-----|
| | | | | | | ENGINE | HOUSE | GENSET |
| | | | | | | | | |
| | | | | | | | | |
| | | | | | | | | |
| | | | | | | | | |
| | | | | | | | | |
| | | | | | | | | |
| | | | | | | | | |
| | | | | | | | | |
| | | | | | | | | |
| | | | | | | | | |
| | | | | | | | | |
| | | | | | | | | |
| | | | | | | | | |
| | | | | | | | | |
| | | | | | | | | |
| | | | | | | | | |
| | | | | | | | | |
| | | | | | | | | |
| | | | | | | | | |
| | | | | | | | | |
| | | | | | | | | |
| | | | | | | | | |
| | | | | | | | | |
| | | | | | | | | |
| | | | | | | | | |
| | | | | | | | | |
| | | | | | | | | |
| | | | | | | | | |
| | | | | | | | | |
| | | | | | | | | |
| | | | | | | | | |
| | | | | | | | | |
| | | | | | | | | |
| | | | | | | | | |
| | | | | | | | | |
| | | | | | | | | |
| | | | | | | | | |
| | | | | | | | | |
| | | | | | | | | |
| | | | | | | | | |
| | | | | | | | | |
| | | | | | | | | |
| | | | | | | | | |
| | | | | | | | | |
| | | | | | | | | |
| | | | | | | | | |
| | | | | | | | | |
| | | | | | | | | |
| | | | | | | | | |

# ENGINE MAINTENANCE

| DATE | ENGINE | HOURS | OIL | TRANS | COOLANT | BATTERIES | | |
|------|--------|-------|-----|-------|---------|-----------|-----|--------|
| | | | | | | ENGINE | HOUSE | GENSET |
| | | | | | | | | |
| | | | | | | | | |
| | | | | | | | | |
| | | | | | | | | |
| | | | | | | | | |
| | | | | | | | | |
| | | | | | | | | |
| | | | | | | | | |
| | | | | | | | | |
| | | | | | | | | |
| | | | | | | | | |
| | | | | | | | | |
| | | | | | | | | |
| | | | | | | | | |
| | | | | | | | | |
| | | | | | | | | |
| | | | | | | | | |
| | | | | | | | | |
| | | | | | | | | |
| | | | | | | | | |
| | | | | | | | | |
| | | | | | | | | |
| | | | | | | | | |
| | | | | | | | | |
| | | | | | | | | |
| | | | | | | | | |
| | | | | | | | | |
| | | | | | | | | |
| | | | | | | | | |
| | | | | | | | | |
| | | | | | | | | |
| | | | | | | | | |
| | | | | | | | | |
| | | | | | | | | |
| | | | | | | | | |
| | | | | | | | | |
| | | | | | | | | |
| | | | | | | | | |
| | | | | | | | | |
| | | | | | | | | |

# ENGINE MAINTENANCE

| DATE | ENGINE | HOURS | OIL | TRANS | COOLANT | BATTERIES | | |
|------|--------|-------|-----|-------|---------|-----------|-----|--------|
| | | | | | | ENGINE | HOUSE | GENSET |
| | | | | | | | | |
| | | | | | | | | |
| | | | | | | | | |
| | | | | | | | | |
| | | | | | | | | |
| | | | | | | | | |
| | | | | | | | | |
| | | | | | | | | |
| | | | | | | | | |
| | | | | | | | | |
| | | | | | | | | |
| | | | | | | | | |
| | | | | | | | | |
| | | | | | | | | |
| | | | | | | | | |
| | | | | | | | | |
| | | | | | | | | |
| | | | | | | | | |
| | | | | | | | | |
| | | | | | | | | |
| | | | | | | | | |
| | | | | | | | | |
| | | | | | | | | |
| | | | | | | | | |
| | | | | | | | | |
| | | | | | | | | |
| | | | | | | | | |
| | | | | | | | | |
| | | | | | | | | |
| | | | | | | | | |
| | | | | | | | | |
| | | | | | | | | |
| | | | | | | | | |
| | | | | | | | | |
| | | | | | | | | |
| | | | | | | | | |
| | | | | | | | | |
| | | | | | | | | |
| | | | | | | | | |
| | | | | | | | | |
| | | | | | | | | |
| | | | | | | | | |
| | | | | | | | | |
| | | | | | | | | |
| | | | | | | | | |

# ENGINE MAINTENANCE

| DATE | ENGINE | HOURS | OIL | TRANS | COOLANT | BATTERIES | | |
|------|--------|-------|-----|-------|---------|-----------|-----|--------|
| | | | | | | ENGINE | HOUSE | GENSET |
| | | | | | | | | |
| | | | | | | | | |
| | | | | | | | | |
| | | | | | | | | |
| | | | | | | | | |
| | | | | | | | | |
| | | | | | | | | |
| | | | | | | | | |
| | | | | | | | | |
| | | | | | | | | |
| | | | | | | | | |
| | | | | | | | | |
| | | | | | | | | |
| | | | | | | | | |
| | | | | | | | | |
| | | | | | | | | |
| | | | | | | | | |
| | | | | | | | | |
| | | | | | | | | |
| | | | | | | | | |
| | | | | | | | | |
| | | | | | | | | |
| | | | | | | | | |
| | | | | | | | | |
| | | | | | | | | |
| | | | | | | | | |
| | | | | | | | | |
| | | | | | | | | |
| | | | | | | | | |
| | | | | | | | | |
| | | | | | | | | |
| | | | | | | | | |
| | | | | | | | | |
| | | | | | | | | |
| | | | | | | | | |
| | | | | | | | | |
| | | | | | | | | |
| | | | | | | | | |
| | | | | | | | | |
| | | | | | | | | |
| | | | | | | | | |
| | | | | | | | | |
| | | | | | | | | |

# ENGINE MAINTENANCE

| DATE | ENGINE | HOURS | OIL | TRANS | COOLANT | BATTERIES | | |
|------|--------|-------|-----|-------|---------|-----------|-------|--------|
| | | | | | | ENGINE | HOUSE | GENSET |
| | | | | | | | | |
| | | | | | | | | |
| | | | | | | | | |
| | | | | | | | | |
| | | | | | | | | |
| | | | | | | | | |
| | | | | | | | | |
| | | | | | | | | |
| | | | | | | | | |
| | | | | | | | | |
| | | | | | | | | |
| | | | | | | | | |
| | | | | | | | | |
| | | | | | | | | |
| | | | | | | | | |
| | | | | | | | | |
| | | | | | | | | |
| | | | | | | | | |
| | | | | | | | | |
| | | | | | | | | |
| | | | | | | | | |
| | | | | | | | | |
| | | | | | | | | |
| | | | | | | | | |
| | | | | | | | | |
| | | | | | | | | |
| | | | | | | | | |
| | | | | | | | | |
| | | | | | | | | |
| | | | | | | | | |
| | | | | | | | | |
| | | | | | | | | |
| | | | | | | | | |
| | | | | | | | | |
| | | | | | | | | |
| | | | | | | | | |
| | | | | | | | | |
| | | | | | | | | |
| | | | | | | | | |
| | | | | | | | | |
| | | | | | | | | |
| | | | | | | | | |
| | | | | | | | | |
| | | | | | | | | |
| | | | | | | | | |
| | | | | | | | | |
| | | | | | | | | |
| | | | | | | | | |
| | | | | | | | | |
| | | | | | | | | |
| | | | | | | | | |

# ENGINE MAINTENANCE

| DATE | ENGINE | HOURS | OIL | TRANS | COOLANT | BATTERIES | | |
|------|--------|-------|-----|-------|---------|-----------|---|---|
| | | | | | | ENGINE | HOUSE | GENSET |
| | | | | | | | | |
| | | | | | | | | |
| | | | | | | | | |
| | | | | | | | | |
| | | | | | | | | |
| | | | | | | | | |
| | | | | | | | | |
| | | | | | | | | |
| | | | | | | | | |
| | | | | | | | | |
| | | | | | | | | |
| | | | | | | | | |
| | | | | | | | | |
| | | | | | | | | |
| | | | | | | | | |
| | | | | | | | | |
| | | | | | | | | |
| | | | | | | | | |
| | | | | | | | | |
| | | | | | | | | |
| | | | | | | | | |
| | | | | | | | | |
| | | | | | | | | |
| | | | | | | | | |
| | | | | | | | | |
| | | | | | | | | |
| | | | | | | | | |
| | | | | | | | | |
| | | | | | | | | |
| | | | | | | | | |
| | | | | | | | | |
| | | | | | | | | |
| | | | | | | | | |
| | | | | | | | | |
| | | | | | | | | |
| | | | | | | | | |
| | | | | | | | | |
| | | | | | | | | |

# ENGINE MAINTENANCE

| DATE | ENGINE | HOURS | OIL | TRANS | COOLANT | BATTERIES | | |
|------|--------|-------|-----|-------|---------|-----------|------|--------|
| | | | | | | ENGINE | HOUSE | GENSET |
| | | | | | | | | |
| | | | | | | | | |
| | | | | | | | | |
| | | | | | | | | |
| | | | | | | | | |
| | | | | | | | | |
| | | | | | | | | |
| | | | | | | | | |
| | | | | | | | | |
| | | | | | | | | |
| | | | | | | | | |
| | | | | | | | | |
| | | | | | | | | |
| | | | | | | | | |
| | | | | | | | | |
| | | | | | | | | |
| | | | | | | | | |
| | | | | | | | | |
| | | | | | | | | |
| | | | | | | | | |
| | | | | | | | | |
| | | | | | | | | |
| | | | | | | | | |
| | | | | | | | | |
| | | | | | | | | |
| | | | | | | | | |
| | | | | | | | | |
| | | | | | | | | |
| | | | | | | | | |
| | | | | | | | | |
| | | | | | | | | |
| | | | | | | | | |
| | | | | | | | | |
| | | | | | | | | |
| | | | | | | | | |
| | | | | | | | | |
| | | | | | | | | |
| | | | | | | | | |
| | | | | | | | | |
| | | | | | | | | |
| | | | | | | | | |
| | | | | | | | | |
| | | | | | | | | |
| | | | | | | | | |
| | | | | | | | | |
| | | | | | | | | |

# ENGINE MAINTENANCE

| DATE | ENGINE | HOURS | OIL | TRANS | COOLANT | BATTERIES | | |
|------|--------|-------|-----|-------|---------|-----------|-----|--------|
| | | | | | | ENGINE | HOUSE | GENSET |
| | | | | | | | | |
| | | | | | | | | |
| | | | | | | | | |
| | | | | | | | | |
| | | | | | | | | |
| | | | | | | | | |
| | | | | | | | | |
| | | | | | | | | |
| | | | | | | | | |
| | | | | | | | | |
| | | | | | | | | |
| | | | | | | | | |
| | | | | | | | | |
| | | | | | | | | |
| | | | | | | | | |
| | | | | | | | | |
| | | | | | | | | |
| | | | | | | | | |
| | | | | | | | | |
| | | | | | | | | |
| | | | | | | | | |
| | | | | | | | | |
| | | | | | | | | |
| | | | | | | | | |
| | | | | | | | | |
| | | | | | | | | |
| | | | | | | | | |
| | | | | | | | | |
| | | | | | | | | |
| | | | | | | | | |
| | | | | | | | | |
| | | | | | | | | |
| | | | | | | | | |
| | | | | | | | | |
| | | | | | | | | |
| | | | | | | | | |
| | | | | | | | | |
| | | | | | | | | |
| | | | | | | | | |
| | | | | | | | | |
| | | | | | | | | |
| | | | | | | | | |
| | | | | | | | | |
| | | | | | | | | |
| | | | | | | | | |
| | | | | | | | | |

# ENGINE MAINTENANCE

| DATE | ENGINE | HOURS | OIL | TRANS | COOLANT | BATTERIES | | |
|------|--------|-------|-----|-------|---------|-----------|-----|--------|
| | | | | | | ENGINE | HOUSE | GENSET |
| | | | | | | | | |
| | | | | | | | | |
| | | | | | | | | |
| | | | | | | | | |
| | | | | | | | | |
| | | | | | | | | |
| | | | | | | | | |
| | | | | | | | | |
| | | | | | | | | |
| | | | | | | | | |
| | | | | | | | | |
| | | | | | | | | |
| | | | | | | | | |
| | | | | | | | | |
| | | | | | | | | |
| | | | | | | | | |
| | | | | | | | | |
| | | | | | | | | |
| | | | | | | | | |
| | | | | | | | | |
| | | | | | | | | |
| | | | | | | | | |
| | | | | | | | | |
| | | | | | | | | |
| | | | | | | | | |
| | | | | | | | | |
| | | | | | | | | |
| | | | | | | | | |
| | | | | | | | | |
| | | | | | | | | |
| | | | | | | | | |
| | | | | | | | | |
| | | | | | | | | |
| | | | | | | | | |
| | | | | | | | | |
| | | | | | | | | |
| | | | | | | | | |
| | | | | | | | | |
| | | | | | | | | |
| | | | | | | | | |
| | | | | | | | | |
| | | | | | | | | |
| | | | | | | | | |
| | | | | | | | | |
| | | | | | | | | |
| | | | | | | | | |

# SHIP'S MAINTENANCE

| DATE | FROM | TO | WORK PERFORMED |
|------|------|----|----------------|
|      |      |    |                |
|      |      |    |                |
|      |      |    |                |
|      |      |    |                |
|      |      |    |                |
|      |      |    |                |
|      |      |    |                |
|      |      |    |                |
|      |      |    |                |
|      |      |    |                |
|      |      |    |                |
|      |      |    |                |
|      |      |    |                |
|      |      |    |                |
|      |      |    |                |
|      |      |    |                |
|      |      |    |                |
|      |      |    |                |
|      |      |    |                |
|      |      |    |                |
|      |      |    |                |
|      |      |    |                |
|      |      |    |                |
|      |      |    |                |
|      |      |    |                |
|      |      |    |                |
|      |      |    |                |
|      |      |    |                |
|      |      |    |                |
|      |      |    |                |
|      |      |    |                |
|      |      |    |                |
|      |      |    |                |
|      |      |    |                |
|      |      |    |                |
|      |      |    |                |
|      |      |    |                |
|      |      |    |                |
|      |      |    |                |
|      |      |    |                |

# SHIP'S MAINTENANCE

| DATE | FROM | TO | WORK PERFORMED |
|------|------|----|----------------|
|      |      |    |                |
|      |      |    |                |
|      |      |    |                |
|      |      |    |                |
|      |      |    |                |
|      |      |    |                |
|      |      |    |                |
|      |      |    |                |
|      |      |    |                |
|      |      |    |                |
|      |      |    |                |
|      |      |    |                |
|      |      |    |                |
|      |      |    |                |
|      |      |    |                |
|      |      |    |                |
|      |      |    |                |
|      |      |    |                |
|      |      |    |                |
|      |      |    |                |
|      |      |    |                |
|      |      |    |                |
|      |      |    |                |
|      |      |    |                |
|      |      |    |                |
|      |      |    |                |
|      |      |    |                |
|      |      |    |                |
|      |      |    |                |
|      |      |    |                |
|      |      |    |                |
|      |      |    |                |
|      |      |    |                |
|      |      |    |                |
|      |      |    |                |
|      |      |    |                |
|      |      |    |                |

# SHIP'S MAINTENANCE

| DATE | FROM | TO | WORK PERFORMED |
|------|------|-----|----------------|
|      |      |     |                |
|      |      |     |                |
|      |      |     |                |
|      |      |     |                |
|      |      |     |                |
|      |      |     |                |
|      |      |     |                |
|      |      |     |                |
|      |      |     |                |
|      |      |     |                |
|      |      |     |                |
|      |      |     |                |
|      |      |     |                |
|      |      |     |                |
|      |      |     |                |
|      |      |     |                |
|      |      |     |                |
|      |      |     |                |
|      |      |     |                |
|      |      |     |                |
|      |      |     |                |
|      |      |     |                |
|      |      |     |                |
|      |      |     |                |
|      |      |     |                |
|      |      |     |                |
|      |      |     |                |
|      |      |     |                |
|      |      |     |                |
|      |      |     |                |
|      |      |     |                |
|      |      |     |                |
|      |      |     |                |
|      |      |     |                |
|      |      |     |                |
|      |      |     |                |
|      |      |     |                |
|      |      |     |                |
|      |      |     |                |

# SHIP'S MAINTENANCE

| DATE | FROM | TO | WORK PERFORMED |
|------|------|-----|----------------|
|      |      |     |                |
|      |      |     |                |
|      |      |     |                |
|      |      |     |                |
|      |      |     |                |
|      |      |     |                |
|      |      |     |                |
|      |      |     |                |
|      |      |     |                |
|      |      |     |                |
|      |      |     |                |
|      |      |     |                |
|      |      |     |                |
|      |      |     |                |
|      |      |     |                |
|      |      |     |                |
|      |      |     |                |
|      |      |     |                |
|      |      |     |                |
|      |      |     |                |
|      |      |     |                |
|      |      |     |                |
|      |      |     |                |
|      |      |     |                |
|      |      |     |                |
|      |      |     |                |
|      |      |     |                |
|      |      |     |                |
|      |      |     |                |
|      |      |     |                |
|      |      |     |                |
|      |      |     |                |
|      |      |     |                |
|      |      |     |                |
|      |      |     |                |
|      |      |     |                |
|      |      |     |                |
|      |      |     |                |
|      |      |     |                |
|      |      |     |                |
|      |      |     |                |
|      |      |     |                |

# SHIP'S MAINTENANCE

| DATE | FROM | TO | WORK PERFORMED |
|------|------|----|----|
|  |  |  |  |
|  |  |  |  |
|  |  |  |  |
|  |  |  |  |
|  |  |  |  |
|  |  |  |  |
|  |  |  |  |
|  |  |  |  |
|  |  |  |  |
|  |  |  |  |
|  |  |  |  |
|  |  |  |  |
|  |  |  |  |
|  |  |  |  |
|  |  |  |  |
|  |  |  |  |
|  |  |  |  |
|  |  |  |  |
|  |  |  |  |
|  |  |  |  |
|  |  |  |  |
|  |  |  |  |
|  |  |  |  |
|  |  |  |  |
|  |  |  |  |
|  |  |  |  |
|  |  |  |  |
|  |  |  |  |
|  |  |  |  |
|  |  |  |  |
|  |  |  |  |
|  |  |  |  |
|  |  |  |  |
|  |  |  |  |
|  |  |  |  |
|  |  |  |  |
|  |  |  |  |
|  |  |  |  |
|  |  |  |  |
|  |  |  |  |

# SHIP'S MAINTENANCE

| DATE | FROM | TO | WORK PERFORMED |
|------|------|-----|----------------|
| | | | |
| | | | |
| | | | |
| | | | |
| | | | |
| | | | |
| | | | |
| | | | |
| | | | |
| | | | |
| | | | |
| | | | |
| | | | |
| | | | |
| | | | |
| | | | |
| | | | |
| | | | |
| | | | |
| | | | |
| | | | |
| | | | |
| | | | |
| | | | |
| | | | |
| | | | |
| | | | |
| | | | |
| | | | |
| | | | |
| | | | |
| | | | |
| | | | |
| | | | |
| | | | |
| | | | |
| | | | |
| | | | |
| | | | |
| | | | |
| | | | |
| | | | |
| | | | |
| | | | |
| | | | |
| | | | |
| | | | |
| | | | |

# IMPORTANT SHIP'S FACTS

Boat Name:

Home Port:

Owners Name(s):

Registration Number:

Documentation Number:

Manufacturer:

Model:

Year:

Hull Number:

Engine:

Engine Serial Number:

Dinghy Engine Fuel Mix:

Dinghy Engine Water Pump:

Dinghy Engine Spark Plug(s):

Dinghy Engine Spark Plug Gap:

Dinghy Engine Point Set:

Dinghy Engine Point Set Gap:

Engine Fuel Filter (Primary):

Engine Fuel Filter (Secondary):

Genset Fuel Filter (Primary):

Genset Fuel Filter (Secondary):

Type & Weight Of Engine/Genset Oil:

Notes:

Dinghy Manufacturer:

Dinghy Registration:

Dinghy Model:

Dinghy Year:

Dinghy Motor Manufacturer:

Dinghy Motor Model Number:

Dinghy Motor Serial Number:

Dinghy Motor Year:

Genset:

Genset Serial Number:

Marine Gear Oil:

Engine Oil Filter:

Genset Oil Filter:

Fuel Tank(s) Capacity:

Water Tank(s) Capacity:

Engine Impeller:

Genset Impeller:

Other Impeller:

Other Impeller:

Other Impeller:

Stuffing Box Packing Size:

# PERIODIC MAINTENANCE ITEMS

Engine Oil/Filter Change:

Genset Oil/Filter Change:

Engine Impeller Change:

Genset Impeller Change:

Marine Gear Oil Change:

Other Impeller Change:

Other Impeller Change:

Other Impeller Change:

Primary Engine Fuel Filter Change:

Secondary Engine Fuel Filter Change:

Primary Genset Fuel Filter Change:

Propane System Tested:

Bilge Pumps Tested:

Rigging Inspected:

Ground Tackle Inspected:

Safety (Pfd, Harness, etc.) Gear Inspected:

Fire Extinguishers Inspected:

Distress Signaling Devices Inspected:

Fuel System Inspected:

Electrical System Inspected:

Water System Inspected:

Waste System Inspected:

Notes:

Secondary Genset Fuel Filter Change:

Shaft Packing Change:

Rudder Stock Packing Change:

Shaft Zinc(s) Change:

Other Zinc(s) Change:

Other Zinc(s) Change:

Other Zinc(s) Change:

Sea Cocks Greased:

Steering Fluid/Cables Checked:

Heat Exchanger Flush:

Coolant Change:

Condition:

Condition:

Condition:

Condition:

Condition:

Condition:

Condition:

Condition:

Condition:

Condition:

Condition:

# LOCKER CONTENTS
## Forward Cabin

Locker Number

1

2

3

4

5

6

7

8

9

10

11

## Galley

Locker Number

1

2

3

4

5

6

7

8

9

10

11

# LOCKER CONTENTS
## Saloon

Locker Number

1

2

3

4

5

6

7

8

9

10

11

## Aft Cabin

Locker Number

1

2

3

4

5

6

7

8

9

10

11

Printed in the United Kingdom
by Lightning Source UK Ltd.
128210UK00001B/18/A